Elderly Care in India

S. Irudaya Rajan · Gayathri Balagopal
Editors

Elderly Care in India

Societal and State Responses

Springer

Editors
S. Irudaya Rajan
Centre for Development Studies
Thiruvananthapuram, Kerala
India

Gayathri Balagopal
Independent Researcher
Chennai, Tamil Nadu
India

ISBN 978-981-10-9874-1 ISBN 978-981-10-3439-8 (eBook)
DOI 10.1007/978-981-10-3439-8

Softcover reprint of the hardcover 1st edition 2017

Printed on acid-free paper

This Springer imprint is published by Springer Nature
The registered company is Springer Nature Singapore Pte Ltd.
The registered company address is: 152 Beach Road, #22-06/08 Gateway East, Singapore 189721, Singapore

Acknowledgements

The senior editor of this volume, as President of the Association of Gerontology (AGI), organized the 17th Biennial Conference of the Association of Gerontology, India (AGI), and International Conference on Engaging and Empowering the Elderly (ICEEE-2014) during September 15–16, 2014, at the Centre for Development Studies, Thiruvananthapuram, Kerala. About 200 scholars and students participated in the conference, and 155 papers were presented both in the plenary and in the parallel sessions such as bio-gerontology, geriatric medicine, psycho-gerontology, elder abuse, economics of aging, health care, social support networks, living with dignity, well-being, living arrangements, and care for the elderly. The seminar was sponsored by HelpAge India; Ministry of Rural Development, Government of India; Ministry of Social Justice and Empowerment, Government of India; and United Nations Fund for Population Activities, New Delhi.

As a chairman of the conference, the senior editor asked some of the participants to take a lead in producing edited volumes based on the papers presented at the conference as we received many good quality papers—rather unusual in most big conferences. We are pleased to mention that two more volumes are published/forthcoming with Springer:

Topics in Bio-medical Gerontology—Editors: P.C. Rath, R. Sharma and S. Prasad, 2017, http://www.springer.com/in/book/9789811021541.

Abuse and Neglect of the Elderly in India—Editors: Mala Kapur Shankardass and S. Irudaya Rajan (forthcoming).

This volume is based on the papers presented in the two themes: care of the elderly and living arrangements. To fill gaps in discussions, we invited more papers. We wish to thank all the contributors who have put in considerable work to bring out the volume on time. We also would like to thank the Springer team at India, in particular Shinjini Chatterjee, Priya Vyas, and K.M. Govardhana for their hard work and encouragement to bring the book out on time.

S. Irudaya Rajan
Gayathri Balagopal

Contents

Editors and Contributors

About the Editors

S. Irudaya Rajan is a Professor at the Centre for Development Studies, Kerala, India. Currently, he is the President of the Association of Gerontology (AGI) as well as Kerala Economic Association (KEA). He has published extensively in national and international journals on social, economic, health and demographic facets of ageing. He is the lead author of the book, *India's Elderly: Burden or Challenge?* (1999), and has edited *Social Security for the Elderly: Experiences from South Asia* (2008). He has developed the first longitudinal Ageing Survey in India (Kerala) since 2004, repeated every 3 years, entering into the fifth wave in 2016. Professor Rajan has been involved in several projects on ageing funded by the United Nations (UN), United Nations Fund for Population Activities (UNFPA), the World Bank, International Labour Organization (ILO), HelpAge International, South Asian Network of Economic Institutes, Indo-Dutch Programme on Alternatives in Development, Shastri Indo-Canadian Institute and International Development Research Centre. He has also undertaken considerable research on migration and is the editor the annual series, India Migration Report as well as editor-in-chief of the journal *Migration and Development.*

Gayathri Balagopal is an independent researcher based in south India. She has worked on health and care needs of the elderly. Her work experience has been in the development sector as researcher with the Banyan Academy of Leadership in Mental Health, Chennai, and as consultant (Research, Monitoring and Evaluation) with the Health Portfolio of the Navajbai Ratan Tata Trust and Sir Ratan Tata Trust, Mumbai. She has researched rehabilitation of homeless women with mental health issues, employment problems confronting persons with mental health issues and documented community mental health programmes in India. Her research interests are in gerontology, public health, mental health and social security.

Contributors

C. Aruna Department of Sociology, Pondicherry University, Puducherry, India

U.R. Arya Centre for Development Studies, Thiruvananthapuram, Kerala, India

Gayathri Balagopal Chennai, Tamil Nadu, India

Dhananjay W. Bansod Department of Public Health and Mortality Studies, International Institute for Population Studies, Deonar, Mumbai, Maharashtra, India

Bharti Birla International Labour Organization, DWT for South Asia and Country Office for India, New Delhi, India

Barnali Das Department of Anthropology, Gauhati University, Guwahati, Assam, India; Department of Anthropology, North Gauhati College, Guwahati, India

Anupama Datta Policy Research and Development, HelpAge India, New Delhi, India

Debasis Barik National Council of Applied Economic Research, New Delhi, India

Deblina Dey Jindal Global Law School, O.P. Jindal Global University, Sonipat, India; Centre for the Study of Law and Governance, Jawaharlal Nehru University, New Delhi, India

Sangeeta Kumari Gupta International Institute for Population Sciences, Longitudinal Ageing Study in India (LASI), Mumbai, Maharashtra, India

S. Irudaya Rajan Centre for Development Studies, Thiruvananthapuram, Kerala, India

S. Jayashree Department of Sociology, Karnatak University, Dharwad, Karnataka, India

A.K. Joshi Department of Sociology, Banaras Hindu University, Varanasi, UP, India

Robert Mathew Mookambika Institute for Medical Sciences, Kulasekharam, Tamil Nadu, India

Chandana Sarmah Department of Anthropology, Gauhati University, Guwahati, Assam, India

T.V. Sekher International Institute for Population Sciences, Department of Population Policies & Programs, Mumbai, Maharashtra, India

Shubham Department of Sociology, Banaras Hindu University, Varanasi, Uttar Pradesh, India

Sonalde Desai National Council of Applied Economic Research, New Delhi, India; University of Maryland, College Park, USA

Sreerupa Centre for Development Studies, Thiruvananthapuram, Kerala, India

S. Sunitha Centre for Development Studies, Thiruvananthapuram, Kerala, India

Tushar Agrawal Centre for Development Management, Indian Institute of Management, Udaipur, Rajasthan, India

List of Figures

List of Tables

List of Map

Chapter 1
Caring India: An Introduction

S. Irudaya Rajan and Gayathri Balagopal

Ageing is very often construed as a crisis—medical as well as social. Population ageing in India is viewed with alarm in the academic literature and popular discourse, described using terms such as "age quake" and "grey tsunami", with implications for the state and the family in terms of economic transfers, health expenditure and caregiving, thus conjuring up a fearful image. This perception is confined not only to the media and academia, but also gets endorsed by the state, which desperately seeks to contain social welfare expenditure on the emerging constituency of elderly persons. Consequently, while making the pitch for targeted social security, the state simultaneously exhorts the family to care for the elderly, invoking Indian tradition, culture and filial piety, ignoring the fact that in the early twentieth century, (1901) life expectancy at birth was just 23 years and infectious diseases, which did not require prolonged caregiving, were the leading causes of death. Age-associated biological changes have been attributed to a slowdown of homoeostatic mechanisms or a decline of adaptability over time (Evans and Williams 1992). Ageing, in such an approach (the biological approach), is studied at the level of organisms, molecules or cells (Bengtson et al. 2005). Demography views population ageing as the proportionate relationship of age groups, wherein population begins to age when the proportion of the elderly in the total population increases because of a decline in the proportion of children. Alterations in the age structure are caused by declines in fertility and mortality, which, in turn, are

S. Irudaya Rajan (✉)
Centre for Development Studies, Prasanth Nagar,
Medical College P.O, Ulloor, Thiruvananthapuram 695 011,
Kerala, India
e-mail: rajan@cds.ac.in; rajancds@gmail.com

G. Balagopal
D-10 Adhisri, New #16 Kothari Road, Nungambakkam, Chennai 600034,
Tamil Nadu, India
e-mail: balagopalgayathri@gmail.com

S. Irudaya Rajan and G. Balagopal (eds.), *Elderly Care in India*,
DOI 10.1007/978-981-10-3439-8_1

influenced by socio-economic development and state action. These representations of ageing as cellular changes or as changes in the age structure do not offer a complete picture of the condition of the elderly population in their social context, as ageing is a multi-textured process comprising biological, demographic and social aspects. Medicalisation of old age obscures the social conditions of the elderly and cannot cure the social maladies experienced by them (Estes and Associates 2001). Rather than ascribing a negative labelling to the ageing experience per se, it is important to realise the important role played by socio-economic conditions and public policy in determining responses to care and support, which influences quality of life among the elderly. The impact of genetic or intrinsic factors on age-related diseases and disabilities depends to a considerable extent on the level of exposure to environmental, lifestyle or extrinsic factors (Evans and Williams 1992), and will vary for different social and economic situations prevailing throughout the life course. In order to gain a comprehensive perspective of the ageing process, we need to recognise two crucial points:

- Ageing takes place in a socio-economic and cultural context, which influences the nature of problems that the elderly experience.
- Ageing is a process that is shaped by events and conditions throughout the life course of individuals.

In such an approach, which combines both the political economy perspective and the life course perspective on ageing, the interaction of the biological processes with the social context of old people determines their quality of life. Both these components of ageing are not mutually exclusive, as the rate of cellular changes that causes ageing depend on the individual's access to social capabilities and material resources, which in turn are governed by the political and economic structures of society. In this context, it is important to understand the demographics of population ageing, health, disability, living arrangements and economic independence among the elderly in India, all of which, in a way, point to the demand for care from the elderly and response to this from the state and social institutions which are the source of elder care.

Demography of Ageing in India

The extent of population ageing of a region is commonly measured using the proportion of the elderly to the total population. In 2011, there were 103,849,040 elderly, comprising 8.6% of the total population in India (Table 1.1). The elderly constituted 8.8 and 8.1% of the population in the rural and urban areas, respectively, in 2011. Table 1.1 illustrates that Kerala (12.6%), Goa (11.2%), Tamil Nadu (10.4%), Punjab (10.3%) and Himachal Pradesh (10.2%) had the highest proportion of the elderly population in India. The rapid decline in fertility is evident across the country, as all the states, except some north-eastern States and some Union

Territories, had a fairly high proportion of elderly population (more than 7%). Among all the states, Arunachal Pradesh (4.6%), Meghalaya (4.7%), Nagaland (5.2%) and Mizoram (6.3%) had a comparatively lower proportion of elderly persons. In fact, among the north-eastern states, only Tripura (7.9%) and Manipur (7%) had a relatively aged population.

An examination of the data on the proportion of elderly between 1961 and 2011 indicates that there is a consistent upward trend from 1961 (Table 1.2). The information on decadal growth rates reveal that the elderly population has grown at a sharp pace after 2001. The structural composition of the elderly has undergone a change between 1961 and 2011, as the share of young–old has decreased from 65.2% in 1961 to 61.7% in 2011, whereas the share of the old–old has increased from 34.8% in 1961 to 38.3% in 2011. The dependency ratio is just a rough proxy of dependence, as the data on worker–population ratio (WPR) show that a significant percentage of the elderly in India work and not all the population in the working age group participate in the workforce. Nevertheless, the dependency ratio indicates the extent of support that the working age population has to provide to the elderly and children. Child dependency ratio is the ratio of the population below 14 years of age (0–14) to the population in the working age (15–59), and aged dependency ratio is the ratio of the population aged 60 years and above to the working age population. According to Irudaya Rajan et al. (1999), total dependency (child and elderly together) declined consistently from 1961 to 1991 in India, and while child dependency ratio declined over time, the aged dependency ratio has increased over time. The data shows that the child dependency ratio is higher than the aged dependency ratio, though the former decreased and the latter increased between 1961 and 2011 in India. The index of ageing, which is the ratio of the elderly population (60 years and above) to the child population (0–14 years), measures the level of population ageing. It takes into consideration the both ends of the age distribution and captures the effect of fertility decline on population ageing. An index value of 30 and above indicates an ageing population (Chakraborti 2004). According to this measure, the ageing process was slow in India during 1961–2001 but gathered pace between 2001 and 2011.

Information on the sex ratio (number of females per 1000 males) among the elderly is of particular importance, as it reflects the differential mortality between males and females (Dandekar 1996). The gender composition has important policy implications, as the nature of problems experienced by the elderly varies among men and women. A masculine sex ratio at old age mirrors continued discrimination against women. Global demographic trends show that the female-to-male sex ratio increases at old age in developed countries, and this is attributed to longer life expectancy among females at older ages when compared to males (Chakraborti 2004). Studies have predicted that much of the global gains in life expectancy will benefit women during 1990–2020 (Davies 1999). However, a larger proportion of older women are likely to be widows (Yesudian 1999). The data reveals that sex ratios among the elderly were female-dominant in India from 2001. In India, the rural–urban distribution of the elderly shows a rural dominance, although there has been a steady increase in the proportion of elderly residing in urban areas. Literacy

Table 1.1 Number and proportion of elderly population in India, state-wise, 2011

States	Number of elderly			Proportion of elderly		
	Rural	Urban	Total	Rural	Urban	Total
NCT Delhi	27,134	1,120,311	114,7445	6.5	6.8	6.8
Haryana	1,512,891	680,864	2,193,755	9.2	7.7	8.7
Himachal Pradesh	649,292	53,717	703,009	10.5	7.8	10.2
Jammu and Kashmir	651,969	270,687	922,656	7.2	7.9	7.4
Punjab	1,957,710	908,107	2,865,817	11.3	8.7	10.3
Rajasthan	3,923,792	1,188,346	5,112,138	7.6	7.0	7.5
Uttarakhand	676,014	224,795	900,809	9.6	7.4	8.9
Chhattisgarh	1,598,547	405,362	2,003,909	8.2	6.8	7.8
Madhya Pradesh	4,194,606	1,518,710	5,713,316	8.0	7.6	7.9
Uttar Pradesh	12,446,468	2,993,436	15,439,904	8.0	6.7	7.7
Bihar	6,868,186	838,959	7,707,145	7.4	7.1	7.4
Jharkhand	1,832,861	523,817	2,356,678	7.3	6.6	7.1
Odisha	3,439,653	544,795	398,4448	9.8	7.8	9.5
West Bengal	4,896,679	2,845,703	7,742,382	7.9	9.8	8.5
Arunachal Pradesh	56,361	7278	63,639	5.3	2.3	4.6
Assam	1,747,513	331,031	2,078,544	6.5	7.5	6.7
Manipur	131,615	68,405	200,020	6.5	8.2	7.0
Meghalaya	109,520	29,382	138,902	4.6	4.9	4.7
Mizoram	32,496	36,132	68,628	6.2	6.3	6.3
Nagaland	81,285	21,441	102,726	5.8	3.8	5.2
Sikkim	33,200	7552	40,752	7.3	4.9	6.7
Tripura	205,763	83,781	289,544	7.6	8.7	7.9
Goa	65,787	97,708	163,495	11.9	10.8	11.2
Gujarat	2,884,326	1,902,233	4,786,559	8.3	7.4	7.9
Maharashtra	6,969,540	4,137,395	11,106,935	11.3	8.1	9.9
Andhra Pradesh (undivided)	6,108,091	2,170,150	8,278,241	10.8	7.7	9.8
Karnataka	3,897,069	1,893,963	5,791,032	10.4	8.0	9.5
Kerala	2,197,552	1,995,841	4,193,393	12.6	12.5	12.6
Tamil Nadu	4,029,097	3,480,661	7,509,758	10.8	10.0	10.4
Andaman and Nicobar Islands	17,939	7485	25,424	7.6	5.2	6.7
Chandigarh	1098	65,980	67,078	3.8	6.4	6.4
Dadra and Nagar Haveli	8638	5254	13,892	4.7	3.3	4.0
Daman and Diu	3583	7778	11,361	5.9	4.3	4.7
Lakshadweep	1099	4171	5270	7.8	8.3	8.2
Puducherry	36,448	83,988	120,436	9.2	9.8	9.7
All India	73,293,822	30,555,218	103,849,040	8.8	8.1	8.6

Source Census of India, 2011

Table 1.2 Demographic characteristics of elderly population (60 years and above) in India, 1961–2011 (as %)

<table>
<tr><th colspan="2">Demographic characteristics of the elderly</th><th>1961</th><th>1971</th><th>1981</th><th>1991</th><th>2001</th><th>2011</th></tr>
<tr><td colspan="2">1. Proportion</td><td>5.6</td><td>6.0</td><td>6.3</td><td>6.7</td><td>7.5</td><td>8.6</td></tr>
<tr><td colspan="2">2. Decadal growth rate</td><td>–</td><td>33.7</td><td>33.0</td><td>29.7</td><td>25.2</td><td>35.5</td></tr>
<tr><td rowspan="3">3. Age</td><td>Proportion of young–old</td><td>65.2</td><td>65.3</td><td>63.2</td><td>63.6</td><td>61.8</td><td>61.7</td></tr>
<tr><td>Proportion of old–old</td><td>34.8</td><td>34.7</td><td>36.8</td><td>36.4</td><td>38.2</td><td>38.3</td></tr>
<tr><td>Total</td><td>100</td><td>100</td><td>100</td><td>100</td><td>100</td><td>100</td></tr>
<tr><td rowspan="2">4. Dependency ratio</td><td>Aged dependency ratio</td><td>10.6</td><td>11.5</td><td>12.0</td><td>12.2</td><td>13.1</td><td>14.2</td></tr>
<tr><td>Child dependency ratio</td><td>77.0</td><td>80.8</td><td>73.4</td><td>67.2</td><td>62.1</td><td>51.0</td></tr>
<tr><td colspan="2">5. Index of ageing</td><td>13.7</td><td>14.2</td><td>16.4</td><td>18.2</td><td>21.1</td><td>27.9</td></tr>
<tr><td colspan="2">6. Sex ratio (females per 1000 males)</td><td>1000</td><td>938</td><td>960</td><td>930</td><td>1029</td><td>1033</td></tr>
<tr><td rowspan="2">7. Rural urban composition</td><td>Rural</td><td>84.9</td><td>83.4</td><td>81.3</td><td>78.1</td><td>75.0</td><td>70.6</td></tr>
<tr><td>Urban</td><td>15.1</td><td>16.6</td><td>18.7</td><td>21.9</td><td>25.0</td><td>29.4</td></tr>
<tr><td rowspan="2">8. Proportion of literate elderly</td><td>Male</td><td>29.2</td><td>NA</td><td>34.8</td><td>40.6</td><td>52.8</td><td>59.1</td></tr>
<tr><td>Female</td><td>4.3</td><td>NA</td><td>7.9</td><td>12.7</td><td>20.3</td><td>28.5</td></tr>
<tr><td rowspan="2">9. WPR among elderly</td><td>Male</td><td>76.6</td><td>73.8</td><td>65.1</td><td>60.5</td><td>60.3</td><td>60.4</td></tr>
<tr><td>Female</td><td>22.4</td><td>10.5</td><td>14.04</td><td>16.1</td><td>20.9</td><td>23.4</td></tr>
</table>

Source Census of India, various years
Note WPR stands for worker population ratio, NA stands for Not Available

is important, as there is a positive association between older people's control of knowledge or information and their status and treatment (Cattel 1992). Also the generation gap is not as pronounced if the elderly are literate (Chakraborti 2004). Effective literacy rate among the female elderly was lower than that among male elderly by nearly 30 percentage points. There has been an increase in literacy among both elderly males and females between 1961 and 2011, although by 2011, it was 59.1% among males and 28.5% among females.

Cross-country comparisons reveal that worker population ratios (WPR) among the elderly were lower in countries with higher per capita income (Audinarayana 2001; Kinsella and Velkoff 2001). In developed countries, the elderly retire with pensions and hence do not have to remain in the workforce. Studies in developed countries have found a significant negative correlation between the ratio of social security to earnings and work participation of the elderly (Clark and Spengler 1980). In contrast to the developed countries, WPR of the elderly, particularly of males was comparatively higher in developing countries such as India. Elderly women in developing countries also have a higher WPR than their counterparts in developed countries (Kinsella and Velkoff 2001). As a result of factors, such as low literacy and educational attainments, low ownership of assets, past employment in the unorganised sector and inadequate coverage by public and private pensions/social security, the elderly have to work in the absence of a substitute for wage income. It is evident that the participation of the elderly in the workforce, by and large, reflects the extent of their vulnerability. In rural Tamil Nadu, studies have shown that 65% of the elderly worked as they did not have any other means of livelihood, while 14% worked to maintain the family, 2% had to work, as they did not have anyone to help and 1% as they did not want to be dependent on others (Nair 1980). However, in a study conducted by Caldwell et al. (1988) in rural Karnataka, the elderly stated that they continued to work because it was a way of life and not a survival strategy. It has been established that while the WPR of elderly men is influenced by economic stress, that of elderly women is influenced by variables such as social and cultural norms, marital status, number and age of children and opportunities for work (Audinarayana 2001; Radha Devi 1992). The data shows that in India, WPR was 60.4% among elderly males and 23.4% among elderly females. While WPR among elderly males declined between 1961 and 2011 that among elderly females increased between 1971 and 2011, suggesting the increasing vulnerability among elderly women. In the absence of occupational pensions for a vast majority of the elderly, a significant proportion of them continue to work.

Marital Status of the Elderly

Among elderly women, there appears to be an overlap between widowhood and ageing (Dreze 1990). The availability of support, residence patterns and well-being of the elderly are influenced by their marital status (Bagga 1999), although some scholars argue that it is more so for elderly men than women (Grundy 2006). Ageing reinforces problems associated with widowhood, which is particularly significant in the case of elderly women, as a large proportion of them are widowed, unemployed, illiterate and dependent on others for economic support. Studies conducted in a few North Indian states have attested to a decline in the economic status of women on widowhood (Chen and Dreze 1995). Widows in northern India experience segregation and isolation from their natal village due to the practice of patrilocality, wherein after marriage, a woman lives in her husband's home, which is in a different village from that of her natal home (Chen 1998). Further, widowhood entails austere practices with regard to dress and behaviour codes as well as social isolation and violence (Chen 1998). It is evident that tradition and culture have a negative connotation, particularly for women. There is a greater incidence of widowhood among women due to considerable spousal age difference, the differences in male and female life expectancy, and the differences between men and women with regard to remarriage (Gulati and Irudaya Rajan 1990).

Table 1.3 shows that, in India, a larger proportion of the elderly (31.5%) than the general population (4.6%) are widowed. In India, as is the case globally, marital status displays a marked variation by gender. In 2011, the proportion of elderly men who were currently married was 82.1%, which was much higher than the proportion of elderly women who were married (49.6%). Further, while only 14.6% of elderly men were widowers, 47.8% of elderly women were widows. This suggests that the majority of elderly men will receive care from their spouse, as most were currently married, whereas the situation of widowed women would depend on their health and functional abilities, prevalence of filial norms, relationship with adult children and other kin networks, and state policies.

Life Expectancy

Given the increase in life expectancy, the socio-economic context under which the population ages is important. Factors like social security, living arrangements, health care, access to occupational pension and availability of kin and other social networks, mechanisms for care and support influence quality of life among the elderly. The latest data on life expectancy demonstrates increase in longevity from 49.7 years in 1970–75 to 67.5 years in 2009–13, representing an increase by 17.8 years (Fig. 1.1). Life expectancy at 60 years has also increased from 13.8 years in 1970–75 to 17.9 years in 2009–13.

Table 1.3 Percentage distribution of the elderly and general population by marital status, India, 2011

Marital status	Gender	Elderly			General population		
		Rural	Urban	Total	Rural	Urban	Total
Never married	Male	2.7	3.3	2.9	52.5	50.3	51.8
	Female	1.7	2.9	2.0	43.0	40.5	42.2
	Person	2.2	3.1	2.5	47.8	45.6	47.1
Currently married	Male	81.5	83.6	82.1	45.1	47.8	46.0
	Female	50.4	47.7	49.6	49.2	51.3	49.9
	Person	65.7	65.4	65.6	47.1	49.5	47.9
Widowed	Male	15.4	12.7	14.6	2.1	1.6	2.0
	Female	47.4	48.8	47.8	7.3	7.6	7.4
	Person	31.7	31.0	31.5	4.6	4.5	4.6
Divorced/Separated	Male	0.4	0.4	0.4	0.3	0.3	0.3
	Female	0.6	0.6	0.6	0.5	0.6	0.6
	Person	0.5	0.5	0.5	0.4	0.4	0.4
All	Male	100	100	100	100	100	100
	Female	100	100	100	100	100	100
	Person	100	100	100	100	100	100

Source Census of India, 2011

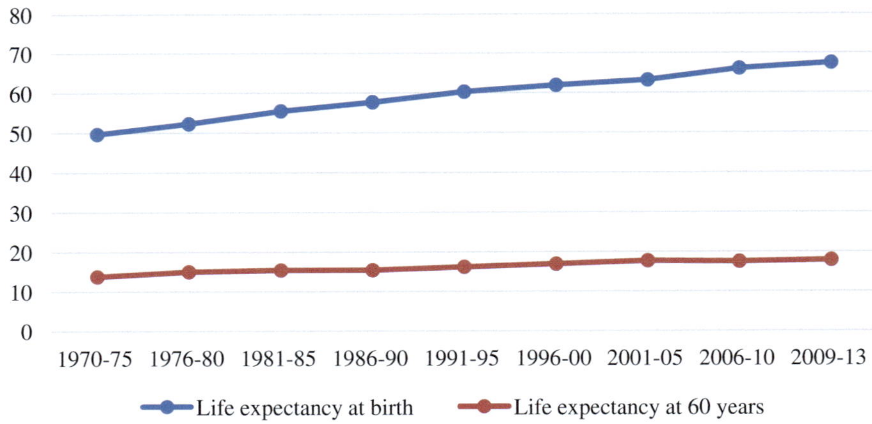

Fig. 1.1 Life expectancy at birth and 60 years, India, 1970–75 to 2009–13 (in years). *Source* Government of India, 2016, Elderly in India

Living Arrangements

The overriding concern in much of the literature has been about the decline of the joint family, which is supposedly caused by the twin forces of urbanisation and modernisation. Cohen (1998) notes that the breakdown of the joint family appears to be the central narrative of Indian gerontology. Field research on family studies has thrown up some important observations regarding the family structure in India. For instance, Shah (1998) points out that jointness was not the norm in traditional Indian society and that it prevailed primarily among the land-owning upper castes. As jointness requires wealth, in the Indian context it was less prevalent among the poor (Van Willigen 2000). Shah (1998) also notes that, over time, urban areas have experienced an increase in incidence of the joint family system, as more often than not married sons cannot afford a place of their own and so continue to live with their parents. Research in Thailand has attributed the high cost of living rather than inter-generational solidarity, as one of the reasons for large households among the poor, and hence well-being of the elderly cannot be judged on the basis of living arrangements alone (Lloyd-Sherlock 2006). A study conducted in New Delhi by Bose (2000) on middle-income group elderly families also corroborates this pattern. Srivastava (2000) argues that the joint family system was not prevalent at all points of time and that the glorification of the elderly in traditional societies was more of an assumption. In fact, the emergence of the joint family in India has been attributed to inadequacy of pension and income (Van Willigen 2000).

Moreover, it is not clear whether co-residence with children necessarily implies that well-being of the elderly is taken care of and that they are dependent on their children, as we also cannot rule out the practice of intra-household discrimination against the elderly. Chen and Dreze (1992: WS-88) have described three determinants of intra-household discrimination in the case of widows, viz. "*the return which decision makers might expect to obtain from allocating consumption to different individuals in the household, the perception of what different members of the household deserve to receive and the bargaining power possessed by different household members engaged in a relationship of co-operative conflict*". Microstudies highlight some of the negative aspects of living with children. A study of some North Indian villages found that elder abuse was prevalent (Srivastava 2000). A survey by HelpAge India (2014) revealed that although 77% of the elderly respondents co-reside with their families, 50% of the elderly had experienced abuse and the main perpetrators were the daughters-in-law (61%) and sons (59%). Recent reports in the media have drawn attention to the traditional practice of geronticide, known as *Thalaikoothal*, in Madurai, Virudhunagar and Theni districts in Tamil Nadu (The Hindu 2016; The Week 2015). This practice prevails even now and has been attributed to poverty and poor access to old age pension. Family members justify this by arguing that they really had no choice due to the declining physical and mental health of their parents and their own economic deprivation. Rituals and practices similar to this have been reported from other parts of the world as well.

Studies show that the elderly express a desire not to co-reside with children but to stay near them, as they perceived that a joint family could cause conflict between the generations owing to higher levels of education, economic independence, and changing values of the younger children (Kishore 2000). A number of factors influence living arrangements, namely marital status, material well-being, health status, family size and structure, cultural traditions such as kinship patterns, the value placed on living independently, the availability of social services and social support and the physical features of housing stock and local communities (Kinsella and Velkoff 2001).

Contrary to prophecies about the decline in shared residential arrangements, and despite increase in urbanisation, there was an increase in the proportion of elderly living with their spouse and adult children between 2004 and 2014, and shared living of the elderly with their family is the dominant pattern in rural and urban India (Table 1.4). Another interesting pattern is that from 2004 onwards, the proportion of elderly living with their spouse and adult children was higher in urban than rural areas. The urban pattern reflects both the poor economic condition of the elderly, as well as the high cost of accommodation, as a result of which the elderly share a household with their adult children. Further, there was a decline in the proportion of elderly living alone between 2004 and 2014 and a rise in the proportion of elderly living with their spouse. Also, the proportion of elderly who lived alone was higher in rural areas in 2004 and 2014. However, there was a slight decline in the proportion of elderly living with their adult children during this period, and this is probably due to increased life expectancy among elderly men and preference of the elderly to live with their spouse.

Table 1.4 Percentage distribution of the elderly by living arrangements India, 1995–1996, 2004, 2014

Living arrangements	Period	Rural	Urban
Alone	1995–1996	4.3	6.0
	2004	5.3	4.3
	2014	4.4	3.5
With spouse only	1995–1996	10.7	5.7
	2004	12.5	10.4
	2014	14.7	15.1
With spouse and adult children	1995–1996	46.2	29.7
	2004	44.2	46.8
	2014	46.4	48.1
With adult children	1995–1996	33.1	51.2
	2004	32.0	32.2
	2014	31.7	30.5
With other relations and non-relations	1995–1996	5.9	6.5
	2004	5.6	4.9
	2014	2.9	2.9
Not recorded	1995–1996	1.0	0.9
	2004	1.9	1.05
	2014	–	–
Total	1995–1996	100	100
	2004	100	100
	2014	100	100

Source National Sample Survey Organisation (1998), 52nd Round, The Aged in India: A Socio-economic Profile, July 1995–June 1996, New Delhi: Government of India; National Sample Survey Organisation, 2006, 60th Round, Morbidity, Health Care and the Condition of the Aged, January–June 2004, New Delhi: Government of India; National Sample Survey Office 2016, 71st Round, Health in India, January–June 2014, New Delhi: Government of India

State of Economic Independence

Microstudies and the literature have identified financial insecurity and poor health as the prime concerns of the elderly (Sen 2000; Irudaya Rajan et al. 1999; Dak and Sharma 1987). Bali (1999) points out that socio-economic vulnerability is greater among the elderly than among younger persons. Agarwal argues that "*inequalities among family members in their ownership endowments, exchange entitlements and access to external support systems would place some members in a weaker bargaining position relative to others. Gender could be one basis of inequality, age another*" (Agarwal 1991). Research has brought out the growing need felt by the elderly to become self-reliant and stresses the importance of savings and government support (Irudaya Rajan et al. 1999; Guruswamy 1988). Increase in the proportion of the elderly could also mean that there will be an increase in the number of the poor elderly (Bose 2000). Economic dependence (45%) and emotional dependence (46%) were one of the main reasons for abuse of the elderly in India (HelpAge India 2014). Treatment of ailments that elderly suffer from will be influenced by their state of economic dependence, as those who provide them with economic support will most likely take the decisions on the utilisation of health care (Rao and Townsend 1998). Increasing age and risk of functional limitations and chronic degenerative diseases diminishes the ability of the elderly to continue working, which leads to cessation of income from employment. At the same, the elderly have to contend with increasing health expenditure, particularly if well-developed, free-of-cost public health services are not available in or near their locality. The combination of limited or zero access to independent sources of income, depleted or no savings and out-of-pocket expenditure on health creates a situation of dependence among the elderly. The economic dependence of the elderly depends on factors such as age, gender, functional capacity, marital status and income (Treas and Logue 1986). The old–old, elderly women, elderly with disabilities, elderly widows and the poor will be dependent on their children, relatives and public assistance. In fact, their survival hinges on various kinship networks. An important point to note in the Indian context is the existence of a legal obligation for adult children to provide economic support to elderly parents under the Maintenance and Welfare of Parents and Senior Citizens Act, 2007, failing which adult children can be penalised by courts if their parents file a case of neglect. NSSO (1998: 4) defines economic independence as "*a person is considered economically independent if he/she does not require to take financial help from others in order to lead a normal life*". The NSSO, however, does not measure the frequency and magnitude of financial transfers to the elderly.

Table 1.5 illustrates that nearly half of the elderly were fully economically dependent on others in rural and urban areas, and there has been a slight decline in this proportion over time. However, the proportion of elderly who were economically independent and partially dependent was higher in the urban than in the rural areas. The data establishes considerable gender differences in the level of economic independence, with a higher degree of economic independence among elderly men (42.6% in rural and 51.5% in urban) than elderly women (11.2% in rural and 12.9% in urban). The sharp gender differentials reflect past participation in the labour market, accumulation of savings and ownership of assets, which are lower among women than men in India. Gender disparities in the level of economic independence will have important consequences for elderly women, as their coping capacities are compromised in the event of declines in health and loss of a spouse.

Despite considerable alarm being raised about the decline in traditional values, inter-generational financial transfers constitute the dominant source of support to the elderly (Table 1.6). Further, the proportion of elderly who were supported by their children increased in rural and urban areas between 1995–96 and 2014. In 2014, children were the source of economic support for nearly 80% of the elderly in

Table 1.5 Percentage distribution of elderly by state of economic independence in India, 1995–96, 2004, 2014

Level of economic independence	Period	Rural			Urban		
		Male	Female	Person	Male	Female	Person
Economically independent	1995–1996	48.5	12.1	30.1	51.5	11.5	31.1
	2004	51.3	13.9	32.7	55.5	17.0	35.9
	2014	42.6	11.2	26.6	51.5	12.9	31.9
Partially dependent	1995–1996	18.0	14.6	16.3	16.9	11.0	13.9
	2004	15.2	12.4	13.8	13.4	9.5	11.4
	2014	23.9	19.1	21.5	18.5	15.3	16.9
Fully dependent	1995–1996	31.3	70.6	51.1	29.7	75.7	53.2
	2004	32.0	72.0	51.9	30.1	72.1	51.6
	2014	33.5	69.6	51.9	30.0	71.8	51.2
Not recorded	1995–1996	2.2	2.6	2.4	1.9	1.8	1.9
	2004	0.2	0.2	0.2	1.0	1.3	1.2
	2014	–	–	–	–	–	–
All	1995–1996	100.0	100.0	100.0	100.0	100.0	100.0
	2004	100.0	100.0	100.0	100.0	100.0	100.0
	2014	100.0	100.0	100.0	100.0	100.0	100.0

Source National Sample Survey Organisation (1998), 52nd Round, The Aged in India: A Socio-economic Profile, July 1995–June 1996, New Delhi: Government of India; National Sample Survey Organisation, 2006, 60th Round, Morbidity, Health Care and the Condition of the Aged, January–June 2004, New Delhi: Government of India; National Sample Survey Office 2016, 71st Round, Health in India, January–June 2014, New Delhi: Government of India

rural and urban areas. This reflects the inadequate coverage and amount of social security pensions. Besides, the increased reliance on adult children has implications for the parent–child relationship, with a higher likelihood of conflict and abuse, particularly if the adult children also live with deprivations.

Health Status of the Elderly

Improvements in life expectancy among the elderly have taken place before commensurate improvements in their socio-economic conditions. The challenge lies in how the elderly will cope with the increased life span and deprivations that most of them, especially women, experience. It is possible that both acute and chronic morbidity will be high among the elderly, although the epidemiologic transition theory implies that there will be a change in the cause of death and disease from communicable to non-communicable because of population ageing. However,

Table 1.6 Percentage distribution of economically dependent aged persons by source of economic support in India, 1995–96, 2004, 2014

Source of economic support	Period	Rural	Urban
Spouse	1995–1996	14.2	15.6
	2004	12.7	14.8
	2014	11.1	15.0
Own children	1995–1996	73.5	72.8
	2004	78.4	76.2
	2014	82.1	80.4
Grand children	1995–1996	5.2	5.5
	2004	2.8	2.6
	2014	2.3	1.2
Others	1995–1996	7.1	6.1
	2004	6.1	6.4
	2014	4.5	3.4
All	1995–1996	100.0	100.0
	2004	100.0	100.0
	2014	100.0	100.0

Source National Sample Survey Organisation (1998), 52nd Round, The Aged in India: A Socio-economic Profile, July 1995–June 1996, New Delhi: Government of India; National Sample Survey Organisation, 2006, 60th Round, Morbidity, Health Care and the Condition of the Aged, January–June 2004, New Delhi: Government of India; National Sample Survey Office 2016, 71st Round, Health in India, January–June 2014, New Delhi: Government of India

when compared to acute morbidity, the incidence of chronic morbidity will be substantially higher among the elderly than among the general population. Gruenberg (1977) has attributed the rise in prevalence of chronic diseases to technical innovations that have prolonged the duration of such illnesses and reduced case-fatality rates. The growing burden of chronic diseases has consequences for the healthcare system as well as the elderly. In India, where healthcare delivery is inadequate, it is doubtful how the system will respond to the increasing prevalence of chronic diseases, particularly among the elderly, i.e. whether the health transition will be accompanied by healthcare transition. Also of importance is the cost of treatment, especially from the point of view of accessibility to the patient. Chronic morbidities, which restrict activities of daily living (ADL) and physical mobility, are the reasons for premature withdrawal from the workforce, which impinges on the well-being of the elderly.

Table 1.7 presents the information on proportion of ailing persons in India. Three distinct patterns can be observed: Firstly, the elderly suffer from higher morbidity levels than the general population. Secondly, it is evident that chronic illness is much higher among the elderly than among the general population. Thirdly, the proportion of elderly who had chronic illnesses was much higher in urban than rural areas.

Table 1.7 Proportion of ailing persons during last 15 days for elderly and general population separately for chronic and short duration ailment, 2014

Type of ailment	Elderly		General population	
	Rural	Urban	Rural	Urban
Chronic	20.4	30.9	4.0	6.7
Short duration	7.2	5.3	4.9	5.1
All	27.6	36.2	8.9	11.8

National Sample Survey Office 2016, 71st Round, Health in India, January–June 2014, New Delhi: Government of India

The data on different types of illnesses among the elderly in India establishes that cardiovascular, musculoskeletal, respiratory, infections and endocrine disorders were higher in rural areas, while cardiovascular and endocrine disorders were higher in urban areas (Table 1.8). Further, the prevalence of these ailments was much higher among the elderly than among the general population. While cardiovascular and endocrine disorders can result in high health expenditure, musculoskeletal ailments can lead to functional limitations and be the cause for premature withdrawal from the work force. These conditions point to the need for economic assistance and social care for the elderly.

Disabilities vary in severity and type, and certain disabilities restrict the functional capacity of the elderly—commonly measured by ADL—which might require care. "*The functional ability of elderly people is crucial to how well they cope with activities of daily living, which in turn affects their quality of life*" (World Health Organisation 1998: 2). The role of families in caregiving assumes importance for the elderly who need assistance to carry out their ADL. Cases of severe disabilities point to the requirement for long-term care (LTC) and rehabilitation services. Policy makers have to consider the elderly population's need for assistive devices such as hearing aids and wheelchairs. It is expected that women will spend more years in a disabled state than men (Suzman et al. 1992).

The data reveals that nearly 5% (5.3%) of the elderly had a disability, and the percentage of the elderly reporting a disability was higher in rural than urban areas among males and females (Table 1.9). Elderly men were more likely to have a disability than women. In addition, disability prevalence was higher among the elderly when compared to the general population.

The percentage of the elderly suffering from visual disability (25.3%) and movement disability (25.3%) was much higher than the proportion with other types of disabilities (Table 1.10). Nearly one-fifth of the elderly reported having a hearing disability.

It would be instructive to examine the extent to which elderly suffer from functional limitations, as this would give a clear indication of the need for care. Table 1.11 reveals that the majority of the elderly were physically mobile in rural and urban areas, and that there was an improvement in this proportion between 2004 and 2014. Further, physical mobility was higher among elderly men than women. Less than 10% of the elderly had functional limitations. Less than 2% of

Table 1.8 Number of ailments of each broad ailment category reported per thousand elderly persons during the last 15 days by gender, 2014

Broad ailment category	Rural		Urban	
	60+	All ages	60+	All ages
1. Infections	37	29	21	27
2. Cancers	1	0	2	1
3. Blood diseases	2	1	3	1
4. Endocrine, metabolic and nutrition	35	8	119	24
5. Psychiatric and neurological	15	5	17	6
6. Eye	8	1	6	2
7. Ear	2	1	1	1
8. Cardiovascular	72	10	142	23
9. Respiratory	39	14	33	18
10. Gastrointestinal	19	7	13	7
11. Skin	4	2	5	3
12. Musculoskeletal	65	12	56	12
13. Genito-urinary	4	2	4	2
14. Obstetric and neonatal	0	0	0	0
15. Injuries	3	2	7	2
16. Other	8	2	15	4
All	314	95	444	132

Source National Sample Survey Office 2016, 71st Round, Health in India, January–June 2014, New Delhi: Government of India

the elderly were confined to the bed (1.7% in rural and 1.5% in urban). Elderly persons who are bedridden would need intense caregiving, as they would require help with all their ADL and might suffer from incontinence, which would aggravate the caregiver's burden. In these cases, the caregiver will face considerable stress, especially in the context of water shortages and limited housing space in urban areas where maintaining hygiene will be a major challenge. Further, without gadgets such as washing machines, and the availability and affordability of sanitary

Table 1.9 Disability prevalence among elderly and general population, India, 2011 (as %)

Gender	Elderly			General population		
	Rural	Urban	Total	Rural	Urban	Total
Male	5.7	4.4	5.3	2.4	2.3	2.4
Female	5.5	4.0	5.0	2.0	2.0	2.0
Person	5.6	4.2	5.2	2.2	2.2	2.2

Source Census of India, 2011

Table 1.10 Distribution of elderly by type of disability, India, 2011 (as %)

Type of disability	Gender	Rural	Urban	Total
Visual	Male	24.3	20.8	23.5
	Female	28.1	23.8	27.1
	Person	26.2	22.3	25.3
Hearing	Male	18.4	19.5	18.7
	Female	18.9	20.9	19.4
	Person	18.7	20.2	19.0
Speech	Male	3.6	6.0	4.2
	Female	2.9	5.0	3.4
	Person	3.3	5.5	3.8
Movement	Male	27.9	26.6	27.6
	Female	23.1	22.5	23.0
	Person	25.5	24.6	25.3
Intellectual disability	Male	1.6	2.2	1.7
	Female	1.5	2.1	1.6
	Person	1.5	2.1	1.7
Mental illness	Male	1.5	2.1	1.6
	Female	1.6	2.4	1.8
	Person	1.5	2.2	1.7
Any other	Male	11.4	14.4	12.1
	Female	9.9	13.3	10.7
	Person	10.6	13.9	11.4
Multiple disability	Male	11.4	8.4	10.7
	Female	13.9	10.1	13.0
	Person	12.7	9.2	11.8
All	Male	100.0	100.0	100.0
	Female	100.0	100.0	100.0
	Person	100.0	100.0	100.0

Source Census of India, 2011

products such as adult diapers, the caregiver will face the tremendous physical burden of doing the laundry manually.

Elder Care in India: Societal and State Responses

Care (formal or informal/paid or unpaid) plays a significant role in the well-being of the elderly (Razavi 2007). According to Razavi (2007), the institutions involved in the provision of care have been conceptualised as a care diamond to include the family/household, markets, the state and non governmental organisations (Fig. 1.2). Factors such as claims of social networks, organised interest groups (for example,

Table 1.11 Percentage distribution of elderly persons of each age group and gender by state of physical mobility, India, 2004 and 2014

State of physical mobility	Period	Rural			Urban		
		Male	Female	Person	Male	Female	Person
Physically mobile	2004	91.2	89.1	90.2	91.7	88.5	90.0
	2014	93.4	89.8	91.6	95.1	90.9	92.9
Confined to bed	2004	1.2	1.4	1.3	1.7	1.7	1.7
	2014	1.2	2.1	1.7	1.0	2.1	1.5
Confined to home	2004	5.5	7.4	6.4	5.1	8.3	6.7
	2014	4.9	7.8	6.4	3.5	6.6	5.1
Move outside the house using a wheelchair	2004	–	–	–	–	–	–
	2014	0.4	0.3	0.3	0.4	0.4	0.4
Not recorded	2004	2.1	2.0	2.1	1.5	1.6	1.5
	2014	–	–	–	–	–	–
All	2004	100.0	100.0	100.0	100.0	100.0	100.0
	2014	100.0	100.0	100.0	100.0	100.0	100.0

National Sample Survey Organisation, 2006, 60th Round, Morbidity, Health Care and the Condition of the Aged, January–June 2004, New Delhi: Government of India; National Sample Survey Office 2016, 71st Round, Health in India, January–June 2014, New Delhi: Government of India

trade unions and women's groups) and state action governs the boundaries of responsibility, which keeps changing (Razavi 2007). Razavi (2007) goes on to argue that the manner in which society deals with care has considerable social significance for gender relations, inequalities and structure of power. Further, the value that society attaches to care should not take the form of interpreting it as something that is performed only by women. "*What is important to note is that the predominance of women in this work is not a result of their free choice or their relative efficiency or inefficiency. The division of work between men and women is largely a social construct—determined by patriarchal traditions and values. In fact, this highly unequal distribution is at the root of power relations between men and women, and all pervasive gender inequalities…unpaid work that contributes*

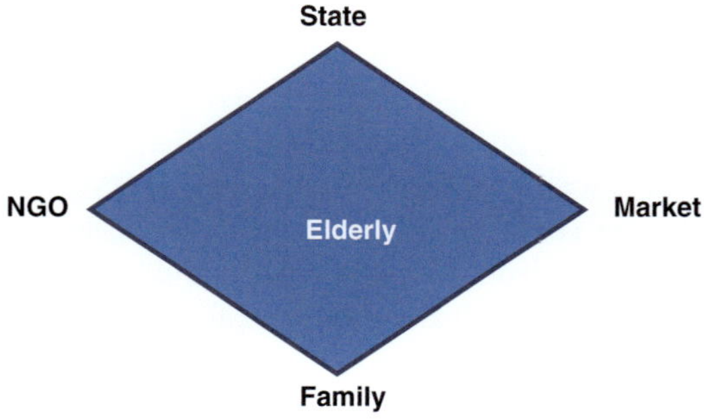

Fig. 1.2 Care diamond. *Source* Razavi (2007: 21)

significantly to the economy is not addressed systematically in policymaking" (Hirway 2015a: 6). She argues that unpaid care work falls outside the boundaries of System of National Accounts (SNA) and non-SNA work, and hence is made invisible in national accounts (Hirway 2015a: 6). Antonopoulos (2009) describes unpaid care work as a kind of social and economic segregation of women from the market and from the mainstream economy. In a very forceful argument about the gendered nature of unpaid care work, Antonopoulos (2009: 2) states, "*The male-breadwinner/female-caregiver polar representation perpetuates a "gendering" ideology that distorts and limits human potential and narrows the range of experiences of "being" and "doing" for men and women. If we are to make further progress towards gender equality, we have to address the fact that it is neither "normal" nor "natural" for women to be performing most of the unpaid labour.... Most importantly, unpaid care work entails a systemic transfer of hidden subsidies to the rest of the economy that go unrecognized, imposing a systematic time-tax on women throughout their life cycle. These hidden subsidies signal the existence of power relations between men and women; also, they connect the "private" worlds of households and families with the "public" spheres of markets and the state in exploitative ways*".

Palriwala and Neetha (2011) examine the care regime through the lens of child care and shed light on the lack of clarity and contradictions inherent in the care regime in India. According to them, "*It is an* ad hoc *summation of informal and stratified practices of (in)security and care. These practices come together on the basis of a fragile institution of family and an ideology of gendered familialism. These frame a discourse that obscures and justifies absences in state actions and governance and tie into a market that is embedded in social and economic inequalities*" (Palriwala and Neetha 2011: 1049). In an analysis of Time Use Survey (TUS) 1998–99 data in India, it emerged that while only 15.8% of men participated in the care of children, sick persons, elderly and persons with disability in the household, 39.9% of women participated in such activities (Neetha 2010). In order to mitigate the disproportionate care burden that women have to bear, Hirway (2015b) suggests that firstly, national accounts have to recognise unpaid work; secondly, care work performed by women has to be reduced with technological and infrastructural support; and thirdly, care work has to be redistributed within the family as well as the other institutions that comprise the care diamond.

Despite India having one of the lowest female labour force participation rates (LFPR), the debate on elder care repeatedly focuses attention on care deficiencies due to women joining paid work. Recent evidence from the NSSO shows that LFPR (usual principal status) among women has declined from 23.7% in 1993–94 to 18.1% in 2011–12 in rural areas and remained almost similar from 13.2 to 13.4% in urban areas during the same period (NSSO 2014a). The decline in LFPR among women could be due to increased demands on them to stay at home, attend to household work and care for the children and the elderly. Findings from another NSSO report illustrate the increase in participation of women in domestic duties

(NSSO 2014b). Among women aged 15–59 years, there has been an increase in the proportion of women participating in domestic duties in rural areas from 53.4% in 2004–05 to 61.6% in 2011–12, and this proportion remains at 65% in 2004–05 and 65.1% in 2011–12 in urban areas (NSSO 2014b). The drop in female LFPR and increased presence of women in domestic duties also suggests that more women will enter old age with financial deprivation, as they lack access to an independent source of income. Besides, as their spouse would have most likely worked in the informal sector, upon widowhood, these women will not receive a family pension.

In India too, the care diamond approach can be used to draw attention to the manner in which care is conceptualised in policy and law and provided by the four institutions, namely the state, family, non-governmental organisations (NGOs) and the market, in addition to the elderly themselves at the centre. The elderly in India have had to develop agency to deal with challenges arising during old age, as a result of inadequate welfare provision by the state and over burdening of the family. Similar to Abe's (2010) description of the changing shape of the care diamond in Japan, in India too, the diamond seems disproportionately focused on two points, namely the family and the elderly themselves. All these institutions are involved in provision of various forms of support and care such as the following:

- Cash transfers
- In-kind transfers such as provision of food grains, cooked food, assistive devices and services such as social care, social contact, residential support such as co-residence with family, geriatric nursing homes, old age homes (OAHs) and day care centres (DCCs).

State Policies

Public policies play an important role in determining social care and economic support provision to the elderly. In countries that follow a welfare-oriented policy, the state complements family care by the provision of important social welfare services to the elderly. According to Cantor (1989), the provision of social care is determined largely by the interaction of state policies with cultural values. For instance, Scandinavian countries provide extensive formal services for the elderly, as the social policies of those countries recognise individual rights and do not coerce adult children to care for their elderly parents (Lowenstein 2005). The interaction between cultural values and public policies influences to some extent the provision of elder care, as is clearly seen in comparative studies of care provision in Europe. Southern European countries such as Italy, Spain and Greece, which are considered to have a familialistic tradition, display higher levels of care from adult children than northern European countries such as Sweden, Denmark and the Netherlands with a welfarist tradition (Ogg and Renaut 2006).

Public policy for the elderly in developing countries has to respond to the care needs of the elderly, while taking into account that old age vulnerabilities are not a product of old age alone, but are the result of the superimposition of random shocks such as illness on deprivations that have been accumulated over the life course. Public policy for the elderly in India does not take into account the social context of old age vulnerability and the heterogeneity of the ageing experience that results from the position of the elderly in the social structure.

Legal Provisions for Economic Support to the Elderly

In developing countries such as India, filial responsibility is an accepted cultural norm and the law mandates financial support towards elderly parents. Section 125 (1)(d) of the Code of Criminal Procedure, 1973 and Section, 20(3) of the Hindu Adoption and Maintenance Act, 1956 recognises the rights of parents without independent means to be supported or maintained by their children who have the financial capacity. The Constitution of India, Article 41 mandates that the state, within its economic capacity, to secure the right to public assistance of the elderly. The Maintenance and Welfare of Parents and Senior Citizens Act was enacted in December 2007 to ensure need-based maintenance for parents and senior citizens and their welfare.

The Act provides for:

- Maintenance of parents/senior citizens by children/relatives made obligatory and justiciable through tribunals
- Revocation of transfer of property by senior citizens in case of negligence by relatives
- Penal provision for abandonment of senior citizens
- Establishment of old age homes for indigent senior citizens
- Protection of life and property of senior citizens
- Adequate medical facilities for senior citizens.

Public Policy on Elder Care: Family as the Diamond

The need for government action arises not because of the erosion of "traditional" value systems or due to modernisation and urbanisation, but because of the failure on the part of the government to provide employment security and universal social security. It is imperative that the government steps in, as markets, are known to fail in social security provision to the needy, and shifting the responsibility to the family and community undermines the role that the government plays in strengthening these institutions (Burgess and Stern 1991).

The National Policy on Older Persons (NPOP), 1999, exhorts families to take care of elderly parents, though it also recognises the importance of supplementing family care with NGO participation (Ministry of Social Justice and Empowerment 1999a). The NPOP encourages the state to consider the needs of the elderly poor in rural and urban areas and directs the state to provide financial security, health care, shelter, welfare schemes, protection against abuse and livelihood opportunities to the elderly. The policy recommends that 10% of the housing schemes meant for the poor be earmarked for the elderly and that social security pensions be revised according to the level of inflation. The importance of social care is recognised, and although the policy considers the family as the primary caregiver of the elderly, it also realises the need to set up voluntary organisations for the provision of old age homes, day care, mobile healthcare services and hospices. Women are generally expected to play the role of the caregiver. Point 11 of NPOP expresses alarm about the changing values in Indian society and states "*Changing roles and expectations of women, their concepts of privacy and space, desire not to be encumbered by caring responsibilities of old people for long periods, career ambitions, and employment outside the home implies considerably reduced time for caregiving*" (Ministry of Social Justice and Empowerment 1999a). Further, point 81 of the policy adds, "*Programmes will be developed to promote family values, sensitise the young on the necessity and desirability of inter-generational bonding and continuity and the desirability of meeting filial obligations. Values of caring and sharing need to be reinforced*". Another initiative, the Old Age and Income Security (OASIS), however, evades financial responsibility on part of the state by stating that subsidies involve a staggering expenditure, which is financially not viable (Ministry of Social Justice and Empowerment 1999b).

Despite reliable evidence from NSSO Reports on the multiple deprivations that the elderly face, the National Policy on Senior Citizens 2011 (NPSC) by the Ministry of Social Justice and Empowerment invokes filial piety and responsibility to centralise care for the elderly within the domain of family and states the policy "*believes in the development of a formal and informal social support system, so that the capacity of the family to take care of senior citizens is strengthened and they continue to live in the family*" (Ministry of Social Justice and Empowerment 2011: 4). The NPSC further bolsters this familialistic orientation by stating that, "*Efforts would be made to strengthen the family system so that it continues to play the role of primary caregiver in old age. This would be done by sensitising younger generations and by providing tax incentives for those taking care of the older members*" (Ministry of Social Justice and Empowerment 2011: 4). Although the available evidence from the NSSO on economic support clearly shows that the family is the dominant financial assistance provider, the NPSC 2011 does not demand state action in the care and support of the elderly. Further, in the context of the low tax base and high informal sector employment in India, state provision of tax breaks for households with elderly members seems out of place.

Unlike in developed countries, there is no public provisioning of home-based social care for the elderly in India. The government runs old age homes and supports NGOs with funds to run old age homes. Institutional care is stigmatised, and the NPSC 2011 states that "*The policy will consider institutional care as the last resort. It recognises that care of senior citizens has to remain vested in the family which would partner the community, government and the private sector*" (Ministry of Social Justice and Empowerment 2011: 4). The NPSC 2011 recognises the need for a higher pension amount (Rs. 1000 per month) for the elderly, which should be indexed to inflation. However, as with most Government of India's schemes, the BPL eligibility criteria remains firmly embedded in the new policy. This last resort approach in India's welfare policy in which the state responds to the welfare needs of those who have no family support, typifies what Palriwala and Neetha (2009: 19) have referred to as family failures, "*which were assumed to be the exception, the pathological as against the socially normal mode of life*". It has been argued that the government puts forth the notion of filial responsibility and home care of the elderly as being more humane to avoid provision of services that families provide otherwise. Literature has referred to this as the "woodwork effect", wherein the government perceives an unprecedented number of people coming out of the woodwork to demand services for their elderly relatives (Moody 2002). In the absence of public social care services, it is the elderly belonging to socioeconomically disadvantaged groups, who will suffer the most, as they will have limited access to paid social care from the private sector owing to their poor purchasing power and consequently become dependent solely on their family.

One of the significant contributions of the state are cash transfers (non-contributory) under the National Social Assistance Programme (NSAP) introduced in 1995. NSAP includes the Indira Gandhi National Old Age Pension Scheme (IGNOAPS), Indira Gandhi National Widow Pension Scheme (IGNWPS), Indira Gandhi National Disability Pension Scheme (IGNDPS), National Family Benefit Scheme (NFBS) and Annapurna. Under the old age pension scheme, the age criteria for the beneficiaries was 65 years and above. The scheme is meant exclusively for destitute elderly people who had no alternate and regular means of subsistence from either their own income or from family. Central assistance towards this scheme was Rs. 75 per month, and the states were allowed to contribute their own share. In 1999, the Government of India also launched a food entitlement scheme called the Annapurna in which the destitute elderly are given 10 kg of rice or wheat every month. The rights of the elderly are considered secondary to the state's fiscal position, and hence social security pension schemes incorporate restrictive eligibility criteria to reduce the number of claimants. Recently, a group of civil society activists have formed a collective called Pension Parishad to campaign for universal old age pension. Currently, old age pension amounts to Rs. 200 per month for those aged 60–79 years and Rs. 500 per month to the oldest (age 80 years or over) who belong to a below-the-poverty line (BPL) household. In addition, the elderly who are eligible also receive an in-kind transfer of 10 kg of food grains every month through the Annapurna Yojana Scheme. Old age pension was first started at Uttar Pradesh in 1957 followed by Kerala in 1960, Andhra Pradesh in 1960 and Tamil Nadu in 1962

(Irudaya Rajan and Mathew 2008). However, there are state-wise variations in the pension amount, with Goa (Rs. 2000 per month), NCT of Delhi, Tamil Nadu and Kerala (Rs. 1000 per month), providing the highest old age pension. Recent data from the India Human Development Survey 2011–12 revealed that 42% of the elderly households in India are covered by old age pension, widows' pension and disability pension, etc. (Barik et al. 2015).

Figure 1.3 illustrates that the percentage of social security expenditure to total expenditure on social services in India between 2005–06 and 2015–16, shows a consistent increase from 2005–06 to 2009–10, following which it stagnated until 2011–12, after which it picked up again, and was 12.8% in 2012–13. After 2012–13, there was a fall in the share of social security expenditure, though there is a revival between 2014–15 and 2015–16.

In 2015, the Government of India launched a contributory pension scheme called Atal Pension Yojana (APY) primarily intended for unorganised sector workers in the age group 18–40 years, who did not receive benefits from any statutory social security scheme. When APY subscribers turn 60 years, they are entitled to receive a fixed pension amount of Rs. 1000 per month, Rs. 2000 per month, Rs. 3000 per month, Rs. 4000 per month and Rs. 5000 per month, depending on the age at which they enrolled and the amount of their contribution. Under the APY, the Government of India will co-contribute 50% of the subscriber's contribution or Rs. 1000 per annum, whichever is lower, guarantee a fixed pension for subscribers and reimburse promotional and development activities that encourage people to enrol for the scheme. The APY imposes penalties in the event of delayed payment and closes the account if premium has not been paid for 24 months. These penalties might work against the subscribers as those in the informal sector and marginal workers might

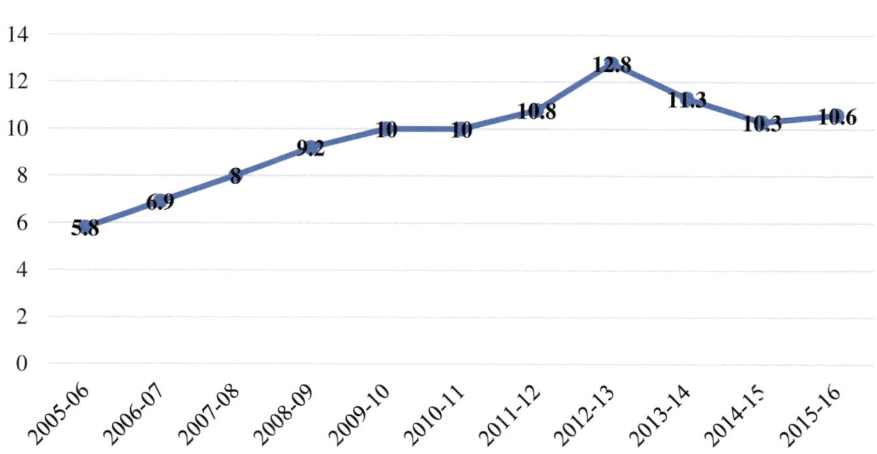

Fig. 1.3 Percentage of expenditure on social security and welfare to social services, India, 2005–06 to 2015–16. *Note* The data for 2014–15 is based on a revised estimate, and for 2015–16 it is based on budget estimates; expenditure includes plan and non-plan expenditure. *Source* RBI, various years, state finances: a study of budgets

not have a regular flow of income. In such situations, subscribers might not be able to meet their premium obligations, leading to drop out from the scheme and deprivation of economic security in their old age. In addition, it remains to be seen whether this programme will result in contracting expenditure on the National Social Assistance Scheme.

Family

Family involvement in elder care is important even in developed countries, though the state plays an equally important role in provision of care services (Lowenstein and Daatland 2006). In a comparative study on elder care in five countries (Norway, England, Germany, Spain and Israel) using the inter-generational solidarity framework, the researchers found that the pattern of exchanges between elderly and adult children was similar across countries with different welfare regimes and family traditions. Families were found to be involved in elder care even in Norway, which is a welfarist state. Lowenstein and Daatland (2006) found that norms about elder care functioned more as guidelines, and actual care provision was determined by situational factors such as competing demands on the family's time. Parental needs such as declining health, functional limitations, increased age and widowhood influenced help. The nature of the inter-personal relationship among parents and adult children affected help to fathers, whereas reciprocity governed care provision to mothers (Lowenstein and Daatland 2006). The inter-generational solidarity framework, which was developed by Bengtson and Roberts (1991), is presented in Table 1.12. The components of the inter-generational solidarity framework were associational solidarity (face-to-face contact), affectual solidarity (emotional closeness, getting along together and communication), consensus (similar values among generations), functional solidarity (social care and economic support), normative solidarity (norms) and structural solidarity (living arrangements).

Both spousal responsibility and filial responsibility are deeply ingrained in most cultures and very often, spousal care is a legal compulsion (Moody 2002). Filial responsibility is governed by a host of factors such as emotional attachment towards the elderly parent, normative solidarity and expectation of financial rewards (Silverstein et al. 1995). Research has shown that the motive for supporting elderly parents varies by the adult child's and the parent's gender. Adult daughters were motivated to care for their elderly parents by affection, whereas adult sons were motivated to do so by social norms. Normative responsibility guides adult sons more than daughters in providing support to elderly parents. The gender of the elderly parent also influences the support provided by adult children. "*Affection is a stronger predictor when mothers are recipients of support, and that inheritance is a more salient predictor when fathers are recipients. Again, these patterns reflect the traditional gender-based division of domains in the family—sentiment inducing support to mothers and financial incentive inducing support to fathers*" (Silverstein et al. 1995: 473). Also, elderly parents with comorbidities and elderly widowed

Table 1.12 Elements of inter-generational solidarity

S. no.	Construct	Nominal definitions	Empirical indicators
1	Associational solidarity	Frequency and patterns of interactions in various types of activities in which family members engage	1. Frequency of inter-generational interactions 2. Types of common activities shared (recreation, special occasions, etc.)
2	Affectual solidarity	Type and degree of positive sentiments held about family members and the degree of reciprocity of these sentiments	1. Ratings of affection, warmth, closeness, understanding, respect, etc. for family members 2. Ratings of perceived reciprocity in positive sentiments among family members
3	Consensual solidarity	Degree of agreement on values, attitudes and beliefs among family members	1. Intra-familial concordance among individual measures of specific values, attitudes and beliefs 2. Ratings of perceived similarity with other family members in values, attitudes and beliefs
4	Functional solidarity	Degree of helping and exchange of resources	1. Frequency of inter-generational exchange of assistance (financial, physical and emotional) 2. Ratings of reciprocity in the intergenerational exchange of resources
5	Normative solidarity	Strength of commitment to performance of familial roles and to meeting familial obligations	1. Ratings of importance of family and inter-generational roles 2. Ratings of strength of filial obligations
6	Structural solidarity	Opportunity structure for inter-generational relationships reflected in the number, type and geographic proximity of family members	1. Residential propinquity of family members 2. Number of family members 3. Health of family members

Source Bengtson and Roberts (1991): 857

mothers were more likely to receive care from both sons and daughters. The gendered nature of elder care has also been documented in a study that found that daughters were more likely than sons to become primary caregivers and that sons assisted elderly parents with support that is culturally perceived to be masculine, such as transportation, home repairs and financial help (Connidis 2005).

Kin networks exert a strong influence on the availability of care and support. The availability of kin depends on the elderly person's family structure and living arrangement. The impact of demographic transition on family structure and living arrangements in societies that have reached replacement level is not hard to imagine. In such societies, there are chances of more beanpole families (fewer numbers in each generation), with adults having more parents than children (Bengtson and Harootyan 1994). The timing of fertility also influences family types—age-condensed and age-gapped families (Bengtson et al. 1995). Age-condensed families are a result of early childbearing and consequently have smaller generational age differences, whereas age-gapped families are the outcome of late childbearing and have larger generational age differences (Bengtson and Harootyan 1994). It has been argued that affective ties between generations will be of a lesser order among age-gapped families (Rossi 1987). A variety of family formats like those who are childless adults, reconstituted families on account of divorce and cohabitation, are also known to affect inter-generational relations (Lowenstein 2005). However, some studies have shown that inter-generational relations remain strong despite changes in the family structure and geographical distance (Bengtson 2000; Lin and Rogerson 1995). In the context of developing countries, where demographic transition is still in the middle stages, age-condensed families are likely to be more common, particularly among the poor because of their early marriage and childbearing patterns. Consequently, more immediate kin are available for elder care, although mere availability does not translate into care.

Examining the ageing experience in urban areas of Indonesia, Eeuwijk (2006) found that the elderly are confronted with a triangle of uncertainty, which has social, economic and health dimensions (Fig. 1.4). This triangle of uncertainty governs formal and informal care for the elderly.

Eeuwijk (2006) highlighted the crux of the issue in Asian countries, namely how appropriate care can be provided to elderly in the context of these uncertainties and inadequate formal care mechanisms. Elder care was found to be provided by kin members, primarily females, such as wives, daughters and daughters-in-law (Eeuwijk 2006). Further, the elderly preferred to carry out their intimate, personal care activities and toileting by themselves. Deterioration in the health of the elderly and a fewer number of caregivers, which increased the burden on available caregivers, were associated with decline in caregiving (Eeuwijk 2006). The risk of inadequate care was higher among widows, unmarried women, the poor and those with vulnerable caregivers (Eeuwijk 2006).

An important feature that characterises social security arrangements in the Indian context is the central role played by the family in acting as a safety net for the elderly in terms of various forms of support such as co-residence, economic support and caregiving.

Living arrangements are known to affect the care received by the elderly, as the availability of a co-resident spouse or adult child might translate into the elderly receiving support on a regular basis for their health and financial needs. The elderly who live alone may experience higher levels of isolation and economic distress, particularly those with disabling health conditions. However, sharing of residence is not a guarantee that the elderly will receive care because of work patterns of their adult children, care responsibilities of adult children towards their own children and intra-family relationships.

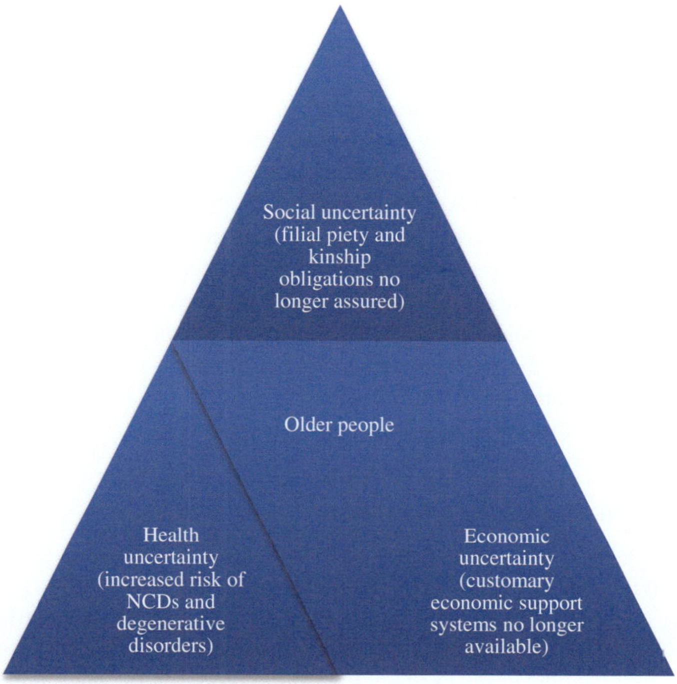

Fig. 1.4 A triangle of uncertainty for elderly. *Source* Eeuwijk (2006: 63)

The capacity of families to take care of their elderly kin is conditioned by competing demands on their time such as caring for young children and employment. Further, absenting from paid employment to care for elderly kin would result in wage loss, as the informal sector does not protect the worker's wages. The need for absenting from work to escort elderly kin to healthcare facilities arises because of the dependence of the elderly belonging to low socio-economic status groups on government healthcare facilities, whose rigid timings conflict with timings at work. This could be an important reason why the poor also show an increased tendency to use private healthcare facilities. Moreover, the cost of treatment in private healthcare facilities could be less than the loss of wages (due to absence from work) incurred if treatment is sought at government health facilities that are free- of- cost or highly subsidised. While work participation of adult children could translate into increased level of economic support to their elderly parents, informal sector employment will not result in increased level of financial assistance to their parents. The provision of economic assistance to the elderly could be curtailed if their adult children have many children, as children under 14 years require their parent's time and economic resources during infancy, schooling, etc.

The kind of relationship shared with the family is also important. For instance, the elderly could have conflicts with their co-resident family members that could negatively affect care and support provision. Adult children who live separately (in the same state, country or in another country) are also important sources of support, depending on the relationship their elderly kin share with them. It is emotional closeness rather than joint living that is the critical factor governing inter-generational relationships (Shanas 1979).

So the extent of social contact maintained with adult children and siblings who live elsewhere are important in establishing a care network for the elderly. Frequent interaction between elderly parents and their adult children and siblings who reside elsewhere helps to maintain the emotional ties among them, which could motivate children and siblings to care for and provide financial support for their elderly kin. Reciprocity is also an important factor affecting the provision of care and support. For instance, family members feel the need to reciprocate for care they received from their parents in the past as well as for their current support in looking after their grandchildren and doing household chores.

Non-governmental Organisations

The government has encouraged NGOs to work in the area of service provision for the elderly population. The Ministry of Social Justice and Empowerment implemented a Central Sector Scheme of Integrated Programmes for Older Persons (IPOP) in 1992 to improve the quality of life of the elderly by providing basic amenities such as shelter, food, medical care and entertainment opportunities, etc., to improve capacity building by Government/Non-Governmental Organizations/Panchayati Raj Institutions/local bodies, etc. In 2014–15, 248 NGOs received assistance under this scheme. The following projects benefit from assistance under the IPOP Scheme:

- Maintenance of Old Age Homes
- Maintenance of Respite Care Homes
- Running of Multi Service Centres for Older Persons
- Mobile Medicare Unit
- Day Care Centre for Care of Old Person with Dementia
- Multifacility Care Centre for Older Widows
- Physiotherapy Clinics
- Regional Resource and Training Centres
- Helplines and Counselling for Older Persons
- Programme for Sensitisation Schools/College Student
- Awareness Projects for Older Persons
- Volunteers Bureau for Older Persons
- Formation of Vridha Sanghas/Senior Citizen Associations/Self Help Groups
- Any other activity, which is considered suitable to meet the objective of the Scheme.

Sawhney (2005) described the role of HelpAge India in developing income-generation activities for the elderly such as candle making, rug weaving and goat rearing as well as micro-credit banks which provide small loans to the elderly.

HelpAge has supported the construction and maintenance of old age homes and day care centres across India (Sawhney 2005). In addition to old age homes, there are NGOs such as ANEW, Tamil Nadu, Red Cross, Kerala, etc., which provide home-based health and social care for the elderly, although these services have to be purchased. So care work at home are redistributed among extra-familial sources, particularly for those families who can afford to pay. Home-based care agencies provide employment opportunities by training women and men from socioeconomically disadvantaged sections of the population to work as nursing aides to care for elderly patients (Razavi 2007). However, there are no regulatory mechanisms to oversee the functioning of these agencies. In addition, there are community-based NGOs, such as Pain and Palliative Care Society in Kerala, which provide free-of-cost, home-based palliative care services by a team of volunteers, including health professionals, for the terminally ill elderly. Many of these NGOs receive funding from the government and corporate sector.

Market

The private sector (for-profit) provides paid home-based health care and social care assistance, as well as nursing home care and old age homes for the elderly who belong to middle- and upper-income groups, which can afford to buy care. Such providers supply tailored home-based medical care packages with professionals such as doctors, nurses, physiotherapist, psychologists and social care services such as nursing aides who assist elderly people in ADL. There has been no comprehensive study of market-based provision of elder care.

Scope of This Work

Elder care is the foci of this work, which attempts to unravel the demographics of ageing, public policy construction of care, and microlevel practice of care. By using a combination of qualitative and quantitative approaches, this volume highlights the range of issues underpinning elder care in India, throwing the spotlight on the challenges that India faces in caring for her elderly with limited state support and near total dependence on the family for social care and economic support. India does not provide universal social security to the elderly, which results in considerable number of older people continuing to work. Those who do not work are economically dependent on their family. Further, the elderly have to contend with the growing burden of ill health and poor functioning, which increases their social and economic dependence on the family, state and non-governmental organisations, and the market. This volume draws on field-based evidence and the legal framework in India to understand the ways in which care is organised for the elderly and the main sources of care provision. This book addresses key themes such as the shrinking of the traditional support base of the elderly, legal aspects of elder care, the need for formal care workers, trajectory of old age homes in India and care arrangements for the elderly in the community.

Chapter 1 by Irudaya Rajan and Balagopal provides a brief overview of the elderly and the care regime in India. Barik et al. examine the three pillars of economic security of the elderly using India Human Development Survey 2 (IHDS 2) data in Chap. 2. While doing so, the chapter identifies the policy challenges and lists suggestions to deal with them. Dey addresses the construction of elder care in Chap. 3. Critically analysing relevant judgments and policies, the chapter examines how the elderly are viewed in policy and law, and how their needs are interpreted. The legal discourse around care in India valorises family care for the elderly. Birla examines the emerging and critical issue of the need for formal care services for the elderly in Chap. 4. When families can afford, they utilise the services of paid care workers, freeing women in the household to take up paid employment. The feminised nature of care work, undervaluation of women's work in general and lack of legislations that recognise and regulate paid domestic work and care work often pushes care workers in precarious employment relationships with poor working conditions and low wages. The demand for paid elderly care, especially long-term care and support, is increasing, but there is an acute shortage of professionally skilled domestic workers who can provide elderly care services. Datta provides an extensive discussion on the role of old age homes in sharing elder care with family in Chap. 5. Traditionally, old age homes were meant for the poor and destitute and were mostly managed by charitable organisations, but in recent years, paid facilities have also emerged to cater to the needs of middle- and upper-middle-class older persons, who can pay for care in old age. However, due to its unstructured growth, there is no regulatory regime to deal with it. This is a cause for concern in more than one way. Irudaya Rajan et al. discuss elder care and living arrangements in Kerala using data from the Kerala Ageing Survey in Chap. 6. The findings of the study revealed that living arrangements of the elderly play a significant role in elderly care. Among older adults who lived with partners, more than 45% were of the view that children are duty-bound to care for their ageing parents. Living arrangements determine the source of care provision when special care is required. Sarmah and Das discuss the socio-economic condition and social support for the ageing Tiwas in Assam in Chap. 7. The joint family is still the dominant living arrangement among them, and the family was the main source of support for the ageing Tiwas. However, they expect that this pattern might not survive in the present form once the younger generation attains higher levels of education. Chapter 8 by Bansod throws light on the perceptions of the elderly in rural Maharashtra on care and support. Results of the study revealed that one-third of the elderly in rural areas felt that they were not getting proper care and support from their children/family, and many of them perceived ageing as a problem. The study highlights the need for the company of either married or unmarried children to make the elderly feel more secure. Shubham and Joshi, using a case study approach, look at caregiving among the oldest-old in Kolkata in Chap. 9. Results indicated that caregiving plays a pivotal role in helping the oldest-old carry out their daily routine activities, as their dependency on caregivers was found to be high in various aspects. In Chap. 10, Sangeeta Kumari and Sekher present their findings on the vulnerability and coping mechanisms of elderly widows in Jharkhand. The study shows that widows had to deal with intra-household and community-level discrimination, had limited or no role in family decision-making, and were economically dependent on others. They suggest strengthening of family and social security measures to take care of elderly widows. Jayashree has examined the caring scenario in the backdrop of feminisation of ageing among old-old widows in Karnataka in Chap. 11. She finds that majority of the old-old widows were independent in their day-to-day activities, but among those who received care and support, sons provided most of the financial support and daughters provided most of the emotional care. Unlike the studies in North India, not all widows in Karnataka viewed widowhood as a problem. Interestingly, many of the elderly widows stated that old age homes were their preferred residential choice. Mathew provides an in-depth discussion on caring for the elderly with dementia in Chap. 12. According to him, caring for patients with dementia is a unique and challenging task, as dementia does not have a cure and takes a protracted course. In addition to supporting the cognitive needs of the patient, caregivers need to adjust to and contain the varied behavioural abnormalities of these patients. Since taking care of people with dementia is a difficult task, the physical and mental health of the caregiver also has to be taken care of. Balagopal studied caregiving arrangements among the elderly living in an urban slum in Tamil State, India, and presents the results in Chap. 13. The findings reveal that most of the elderly were able to take care of their social care needs though they were highly dependent for economic support. Social care displayed gender asymmetry, with women providing the bulk of care to the elderly, whereas most of the care recipients were elderly men. However, elderly women seem to be better resourced in terms of social networks to deal with old age, as they received care from diverse sources, unlike elderly men, whose predominant source of social care was their spouse. Aruna addresses sibling care among rural elderly women in rural Tamil Nadu in Chap. 14. Findings indicate the existence of strong ties between siblings, and that brothers play a dominant role until the children are married. At later life, both brothers and sisters converge as confidants and companions. Sisters tend to form a thicker bond similar to friends and brothers provide the ritual and service support. In Chap. 15, Sreerupa discusses the implications of transnational migration of younger generation on the care for the older persons staying behind, examining the transnational care strategies and provisioning of care within the transnational families, in the context of Kerala, which is at the forefront of both migration and ageing. In the study area, the middle- and upper-middle-class Syrian Christian migrants' access to economic resources, strong social network, integration into the labour market and legal protection in the host country empowered them with a wider set of resources and capabilities to engage in transnational practices like "caring from a distance" as well as purchase of local paid care services.

Future Research

Larger Studies Across States Using a Comparable Framework

The issue of elder care needs to be studied across the country using a comparable framework. The inter-generational solidarity framework developed by Bengtson and Roberts (1991) could be adapted to the Indian situation or a new framework can be developed to study care and support for the elderly. A gender-based perspective, which moves away from the patriarchal structuring of care as a woman's domain, should inform research on elder care. The stress on family care for the elderly without a discussion on men's sharing of care responsibilities only serves to strengthen and perpetuate gender asymmetry in care provision. Studies need to explore the frequency and intensity of care for the elderly. While most of the discourse centres on the norms of filial piety, studies should look at this issue in all the states to understand actual societal expectations about elder care. Marked variations are found in status of women in North and South India, and this should be examined in detail. Further, there is a need to initiate prospective longitudinal studies on inter-generational relationships among the cohorts of elderly, which will throw light on the interaction between socio-economic changes and changes in nature and form of family relationships comparing low fertility rates, particularly in states such as Kerala, Tamil Nadu, Punjab and Goa, where the pool of potential caregivers shrinks with high fertility states such as Assam, Uttar Pradesh and Bihar.

Social Networks

The size, composition and strength of social networks among the elderly have not been explored in detail in the Indian context. Social networks of the elderly represent coping strategies among the elderly to mediate age-related challenges. Of interest are how social networks translate into social support for the elderly and how these networks and support flows change with increase in age.

Caregiver Burden/Perspective

While looking at elder care, it is important to examine the caregiver's perspective. There are very few studies that have looked at caregiver stress and burden. Prolonged caregiving, particularly for bedridden elderly can result in caregiver's burnout due to poor technological and infrastructural support. Caregivers' perspectives on redistribution of care work among other family members need to be studied.

Institutional Care

While there are studies on the quality of life among the elderly who live in OAHs, there is a need to look at how their care needs are met. Besides, the mechanisms of social networks and companionship have to be assessed.

Market and NGO-Provided Home-Based Services

As the issue of elder care is in a nascent stage, elder care services provided by the market and NGOs need attention. There should be a mapping of such facilities and evaluation of the different elder care services that this segment provides. This is important, as the majority of India's workforce is employed in the informal sector with relatively low pay which raises questions about the affordability of paid elder care.

Conditions of Formal Care Workers

The conditions of work among formal care workers or paid care workers such as nursing aides employed by NGOs and the for-profit private sector require investigation. Issues of adequate training to cope with the demands of elder care, particularly among elderly persons who are bedridden or among those diagnosed with dementia, and relations between the care workers and NGO/Private-for-Profit agencies as well as between care workers and families who engage them, need to be understood.

NSSO Surveys on Elderly Should Include Caregiving

At present, the NSSO's surveys on the elderly focus on economic support. In future rounds, the survey should include aspects related to caregiving for the elderly.

Acknowledgements Some parts of this chapter draws on Gayathri Balagopal's unpublished Ph.D. Thesis entitled "Morbidity, Medical Treatment, Social Care and Economic Support of the Elderly: A Case Study among the Urban Poor in Tamil Nadu State, India" (2009) at the Madras Institute of Development Studies, Chennai.

References

Abe, A. K. (2010). The changing shape of the care diamond: The case of child and elderly care in Japan, gender and development programme Paper No. 9. Geneva: UNRISD.

Agarwal, B. (1991). Social security and the family: Coping with seasonality and calamity in rural India. In E. Ahmad, J. Dreze, J. Hills, & A. Sen (Eds.), *Social security in developing countries*. Oxford: Clarendon Press.

Antonopoulos, R. (2009). The unpaid care work—paid work connection. International labour office, policy integration and statistics department. Geneva: ILO. Working Paper No. 86.

Audinarayana, N. (2001). Factors affecting the work participation of the elderly: An empirical investigation. *Demography India, 30*(1), 61–72.

Bagga, A. (1999) Women and health in old age: the Indian scenario. In *Ageing and Health: A Global Challenge for the 21*st *Century, Proceedings of a WHO Symposium, Kobe 10–13 November. 1998* (pp. 318–328). Geneva: World Health Organisation.

Bali, A. (1999). *Understanding greying people of India*. New Delhi: Inter-India.

Barik, D., Agrawal, T., & Desai, S. (2015). After the dividend: Caring for a greying India. *EPW Economic and Political Weekly, 1*(24), 108–112.

Bengston, V. L. (2000). Beyond the nuclear family: The increasing importance of multigenerational bonds. *Journal of Marriage and the Family, 63,* 1–16.

Bengston, V. L., & Harootyan, R. A. (1994). *Intergenerational linkages*. New York: Springer.

Bengtson, V. L., Rosenthal, C. J., Burton, L. M., (1995). "Paradoxes of families and ageing", In R. H. Binstock, & L. K. George (Eds.), *Handbook of Ageing and the Social Sciences*, (4th ed.), (pp.253–282). San Diego, California: Academic Press.

Bengtson, V. L., Putney, N. M., & Johnson, M. L. (2005). The problem of theory in gerontology today. In M. L. Johnson, V. L. Bengston, P. G. Coleman, & T. B. L. Kirkwood (Eds.), *The Cambridge handbook of age and ageing* (pp. 3–20). Cambridge: Cambridge University Press.

Bengtson, V. L., & Roberts, R. E. L. (1991). Intergenerational solidarity in aging families: An example of formal theory construction. *Journal of Marriage and the Family, 53,* 856–870.

Bose, A. (2000). Demographic transition. *Seminar, 488,* 35–39.

Burgess, R., & Stern, N. (1991). Social security in developing countries: What, why, who and how? In E. Ahmad, J. Dreze, J. Hills, & A. Sen (Eds.), *Social security in developing countries*. Oxford: Clarendon Press.

Caldwell, J. C., Reddy, P. H., & Caldwell, P. (1988). *The causes of demographic change: Experimental research in South India*. Wisconsin: The University of Wisconsin Press.

Cantor, M. H. (1989). Social care: Family and community support systems. *Annals of the American Academy of Political and Social Sciences, 503,* 99–112.

Cattel, M. G. (1992). Knowledge and social change in Samia, Western Kenya. In P. Krishnan & K. Mahadevan (Eds.), *The elderly population in developed and developing world: Policies* (pp. 121–150). New Delhi: BR Publishing Corporations, Problems and Perspectives.

Chakraborti, R. D. (2004). *The greying of India: Population ageing in the context of Asia*. New Delhi: Sage.

Chen, M A., & Dreze J. (1995). Widowhood and well-being in rural North India. In M. Dasgupta, L.C. Chen, T.N. Krishnan (Eds.). *Women's health in India: risk and vulnerability* (pp. 245–288). Bombay: Oxford University Press.

Chen, M. (1998). Introduction. In Martha Chen (Ed.), *Widows in India: Social neglect and public action* (pp. 19–59). New Delhi: Sage.

Chen, M., & Dreze, J. (1992). Widows and health in rural North India. *Economic and Political Weekly, 27* (43/44), WS81–WS92.

Clark, R. L., & Spengler, J. J. (1980). *The economics of individual and population aging*. Cambridge: Cambridge University Press.

Cohen, L. (1998). *No aging in India: Alzheimer's, the bad family and other modern things*. Berkely: University of California Press.

Connidis, I. A. (2005). Siblings ties across time: The middle and later years. In M. L. Johnson, V. L. Bengston, P. G. Coleman, & T. B. L. Kirkwood (Eds.), *The Cambridge handbook of age and ageing* (pp. 429–436). Cambridge: Cambridge University Press.

Dak, T. M., & Sharma, M. L. (1987). Changing status of the aged in North Indian villages. In M. L. Sharma & T. M. Dak (Eds.), *Aging in India* (pp. 43–55). New Delhi: Ajanta Publications.

Dandekar, K. (1996). *The elderly in India*. New Delhi: Sage.

Davies, A.M. (1999). Ageing and health in the 21st century: An overview. *In Ageing and Health: A Global Challenge for the 21*st *Century, Proceedings of a WHO symposium, Kobe 10–13, November 1998* (pp. 206–351).

Dreze, J. (1990). Widows in rural India. Discussion Paper No. 26, development economics research programme. United Kingdom: London School of Economics.

Eeuwijk, P. Van. (2006). Old-age vulnerability, ill-health and care support in urban areas of Indonesia. *Ageing and Society, 26,* 61–80.

Estes, C. L, & Associates. (2001). Social policy and aging: A critical perspective. Thousands Oaks: Sage.

Evans, J. G., & Williams, T. F. (1992). Introduction. In J. G. Evans & T. F. Williams (Eds.), *Oxford textbook of geriatric medicine*. Oxford: Oxford University Press.

Gruenberg, E. M. (1977). The failures of success. *Milbank Memorial Fund Quarterly, 55,* 3–24.

Grundy, E. (2006). Ageing and vulnerable elderly people: European perspectives. *Ageing and Society, 26,* 105–134.

Gulati, L., & Irudaya Rajan, S. (1990). Social and economic implications of population ageing in Kerala, India. *Demography India, 19*(2), 235–250.

Guruswamy, M. (1988). The changing concept of old-age security in relation to fertility decline. In K. Srinivasan & S. Mukerji (Eds.), *Dynamics of population and family welfare*. Bombay: Himalaya Publishing House.

Hirway, I. (2015a). Unpaid work and the economy: Linkages and their implications. Working Paper No. 838, Levy Economics Institute of Bard College.

Hirway, I. (2015b). *Need for policy shift: Women with burden of unpaid work take up jobs of poor quality, low wages, poor or no social protection*. COUNTERVIEW: ORG.

HelpAge India. (2014). *Elder abuse in India*. New Delhi: HelpAge India.

http://www.censusindia.gov.in/.

Irudaya Rajan, S., & Mathew, E. T. (2008). India. In S. Irudaya Rajan (Ed.), *2008, Social security for the elderly: Experiences from South Asia* (pp. 39–106). New Delhi: Routledge.

Irudaya Rajan S., Mishra, U. S., & Sarma, P. S. (1999). *India's elderly: Burden or challenge?*. New Delhi: Sage.

Kinsella, K., & Velkoff, V. A. (2001). *An aging world: 2001*. Washington DC: United States Census Bureau.

Kishore, S. (2000). Ageing pains. *Seminar, 488,* 30–34.

Lin, G., & Rogerson, P. A. (1995). Elderly parents and geographic availability of their adult children. *Research on Aging, 17,* 303–331.

Lloyd-Sherlock, P. (2006). Identifying vulnerable older people: insights from Thailand. *Ageing and Society, 26,* 81–103.

Lowenstein, A. (2005). Global ageing and challenge to families. In M. L. Johnson, V. L. Bengston, P. G. Coleman, & T. B. L. Kirkwood (Eds.), *The Cambridge handbook of age and ageing* (pp. 403–412). Cambridge: The Cambridge University Press.

Lowenstein, A., & Daatland, S. A. (2006). Filial norms and family support in a comparative cross-national context: Evidence from the OASIS study. *Ageing and Society, 26,* 203–223.

Ministry of Social Justice and Empowerment. (1999a). *National Policy on Older Persons*. New Delhi: Government of India.

Ministry of Social Justice and Empowerment. (1999b). *First report of project OASIS*. New Delhi: Government of India.

Ministry of Social Justice and Empowerment. (2011). *National policy for senior Citizens*. New Delhi: Government of India.

Moody, H. R. (2002). *Aging: Concepts and controversies* (4th ed.). Thousand Oaks: Sage.

Nair, T. K. (1980). *Older people in rural Tamil Nadu*. Madras: Madras School of Social Work.

National Sample Survey Office. (2014a). 68th round, employment and unemployment situation in India 2011–12, Report No. 554. New Delhi: Government of India.

National Sample Survey Office. (2014b). 68th round, participation of women in specified activities along with domestic duties 2011–12, Report No. 559. New Delhi: Government of India.

National Sample Survey Organisation. (1998). *52nd round, the aged in India: A socio-economic profile, July 1995–June 1996*. New Delhi: Government of India.

National Sample Survey Office. (2016). *71st Round, Health in India, January–June 2014*, New Delhi: Government of India.

Neetha, N. (2010). Estimating unpaid care work: Methodological issues in time use surveys. *Economic & Political Weekly, XLV, 44,* 73–80.

Ogg, J., & Renault, S. (2006). The support of parents in old age by those born during 1945–1954: A European perspective. *Ageing and Society, 26,* 723–743.

Palriwala, R., & Neetha, N. (2009). *The care diamond: State social policy and the market, India: Research Report 3*. Geneva: UNRISD.

Palriwala, R., & Neetha, N. (2011). Stratified familialism: The care regime in india through the lens of childcare. *Development and Change, 42*(4), 1049–1078.

Radha Devi, D. (1992). Work participation rate of the aged in India. In P. Krishnan & K. Mahadevan (Eds.), *The elderly population in developed and developing world: Polices* (pp. 424–443). Problems and Perspectives, New Delhi: BR Publishing Corporation.

Rao, S. R., & Townsend, J. W. (1998). Health needs of elderly Indian women: An emerging issue. In M. Krishnaraj, R. M. Sudarshan, & A. Shariff (Eds.), *Gender* (pp. 340–363). New Delhi: Oxford University Press, Population and Development.

Razavi, S. (2007). The political and social economy of care in a development context: Conceptual issues, research questions and policy options, gender and development programme Paper No. 3. Geneva: UNRISD.

Rossi, A. S. (1987). Parenthood in transition: From lineage to child to self-orientation. In J. Lancaster, J. Altman, A. S. Rossi, & L. Sherrod (Eds.), *Parenting across the life span: Biosocial dimensions* (pp. 31–84). New York: Aldine de Gruyter.

Sawhney, M. (2005). The role of non governmental organisations for the welfare of the elderly. In P. S. Liebig & S. Irudaya Rajan (Eds.). *An aging India: perspectives, prospects and policies* (pp. 179–191). New Delhi: Rawat.

Sen, K. (2000). Demographic ageing and social development. In *Proceedings of International Conference 'Health Systems and Social Development: An Alternative Paradigm in Health Systems Research', Maastricht.*

Shah, A. M. (1998). *The family in India: Critical essays*. Hyderabad: Orient Longman.

Shanas, E. S. (1979). The family as a social support system in old age. *The Gerontologist, 19*(2), 169–174.

Silverston, M., Parrot, T. M., Bengston, V. L. (1995). Factors that predispose middle—aged sons and daughters to provide social support to older parents. *Journal of Marriage and the Family, 57*(2), 465–475.

Srivastava, V. K. (2000). Ageism. *Seminar*, 488, 21–25.

Suzman, R., Kinsella, K. G. & Myers, G. C. (1992). Demography of older population in developed countries. In J. G. Evans & T. F. Williams (Eds.). *Oxford text book of geriatric medicine: Oxford*. Oxford University Press.

The Hindu. (2016). The morbid practice of facilitated deaths in TN, Dennis S Jesudasan, February 18, 2016.

The Week. (2015). Old age poison, Lakshmi Subramaniam, November 22, 2015.

Treas, J., & Logue, B. (1986). Economic development and the older population. *Population and Development Review, 12*(4), 645–673.

Van Willigen, J. (2000). Social ageing in India and America. *Seminar*, 488, 26–29.

World Health Organisation. (1998). *Growing old-staying well: Ageing and physical activity in daily life*. Geneva: World Health Organisation.

Yesudian, C. A. K. (1999). Socioeconomic implications of ageing. In *Ageing and Health: A Global Challenge for the 21st Century, Proceedings of a WHO symposium, Kobe 10–13, November 1998* (pp. 36–44).

Chapter 2
After the Dividend: Caring for a Greying India

Debasis Barik, Tushar Agrawal and Sonalde Desai

Abstract As in any other society, in India too, the economic security of the aged is based on three main sources: their own income and savings, support from the extended family, particularly children, and support from the state. As India moves rapidly towards a demographic future in which the elderly form a large part of the population, this article examines trends in each of the three supports. While doing so, it identifies the policy challenges and lists suggestions to deal with them.

Keywords Economic security · Greying india · Care · Public transfer · Work · Family remittances

After the Dividend: Caring for a Greying India

National euphoria about reaping the demographic dividend has blinded us to the drama that is unfolding almost out of our sights, slow ageing of the Indian society. As we race towards a demographic future in which the elderly form a larger and larger part of population distribution, India is also experiencing transformation in its economic and social life. We take stock of some of these changes and identify policy challenges lurking on the horizon.

Debasis Barik (✉) · Sonalde Desai
National Council of Applied Economic Research, Parisila Bhawan,
11 Indraprastha Estate, New Delhi 110002, India
e-mail: dbarik@ncaer.org

Sonalde Desai
University of Maryland, College Park, USA
e-mail: Sonalde.desai@gmail.com

Tushar Agrawal
Centre for Development Management, Indian Institute of Management,
Udaipur 313001, Rajasthan, India
e-mail: tushar.agrawal@iimu.ac.in

S. Irudaya Rajan and G. Balagopal (eds.), *Elderly Care in India*,
DOI 10.1007/978-981-10-3439-8_2

In 1991, Indian population contained 57 million elderly; this population has now doubled to 104 million (Census of India 2011). Much of this increase is due to overall population increase, but fertility decline has contributed to increase the proportion of the elderly in the Indian population from 6.6% in 1991 to 8.6% in 2011 (Census of India 2011). However, ageing in India is not uniform across states (Fig. 2.1). States that began fertility decline early have a higher proportion of the elderly than states that still show moderate to high fertility.

According to Census of India, 2011, the proportion of elderly in most of the states falls within the range of 7–10%. Only Kerala is at the high end. One in every eight persons in Kerala is over 60. Some other states such as Tamil Nadu, Punjab and Himachal Pradesh are also closely following the same path. Since proportion of elderly in any society is dwarfed by proportion of children until replacement level fertility is achieved, the proportion of elderly is smaller—about 7%—in high-fertility states like Uttar Pradesh, Bihar and Madhya Pradesh, but as these states experience fertility transition, they will also face the burden of an ageing population.

While the demographic scenario is propelling us towards inevitable ageing over the next fifty years, social transformation already under way is shifting the fundamental props of elderly support.

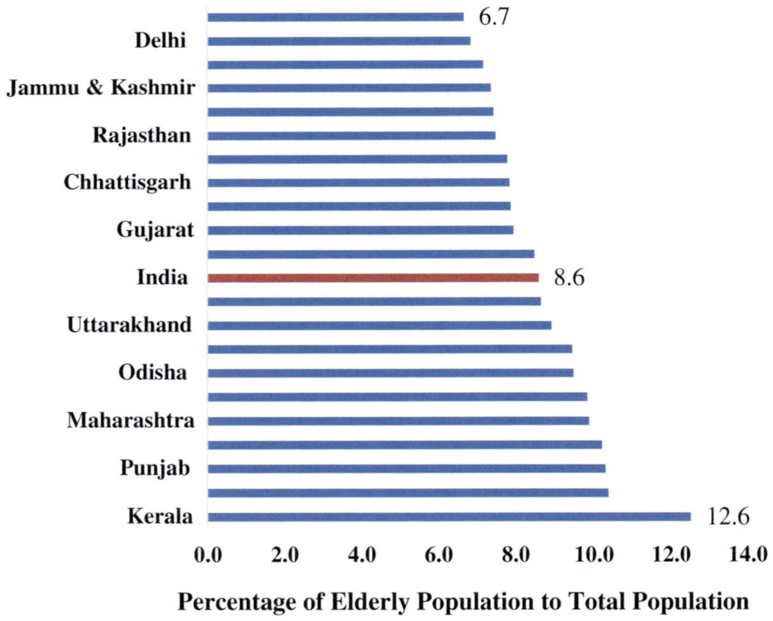

Fig. 2.1 Proportion of elderly population (aged 60 and above) in India, 2011. *Source* Authors' calculations based on Census 2011 data

Income and Consumption Across the Life Cycle

Figure 2.2, drawn from the research on the spread of income and consumption across the life cycle, documents two striking patterns (Lee and Mason 2011): (1) in any given country, the young and the old tend to consume more than they produce; and (2) when we compare across countries, although there are some differences between the rich and the poor countries in age-specific consumption and income patterns, the most striking divergence appears in consumption at the old age.

Comparison Over the Life Course

In the early phase of demographic transition, the aggregate economic life cycle is dominated by the large life cycle deficit (consumption minus production) of the young. The net deficit in early ages is mainly due to the cost related to rearing of the children, and investment on education and skill building activities. This leads to a higher consumption in the absence of any income during these days. Over the course of the demographic transition, population's age and the life cycle deficit of the old become increasingly important. The deficit in the later ages is associated

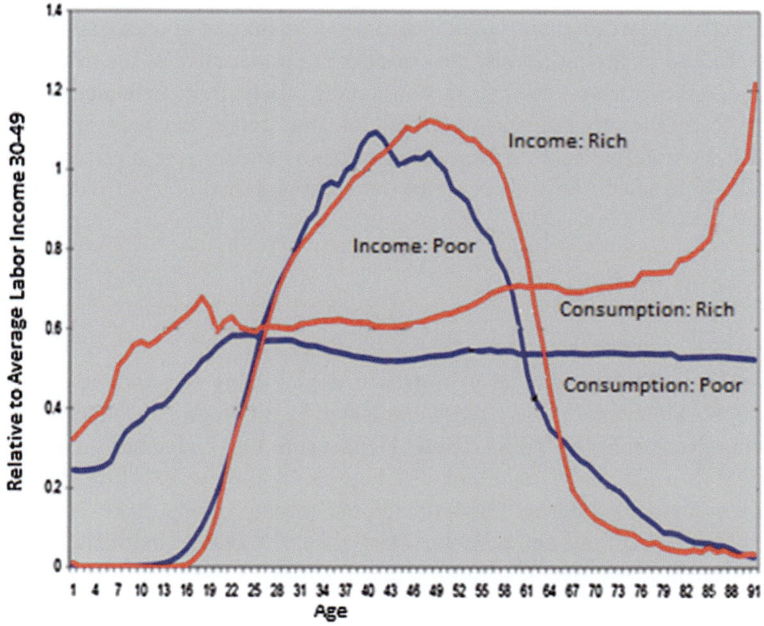

Fig. 2.2 The economic life cycle of rich and poor countries: consumption and labour income. *Source* Lee and Mason (2011, p. 82)

with a higher consumption due to high treatment cost of late life chronic morbidities and lower productivity.

Comparison Across Countries

Though the life cycle levels of income differ substantially between the rich and the poor nations, the patterns seem similar. However, when it comes to consumption, a huge difference can be observed between the rich countries like the USA, Japan, Finland and Sweden and the poor countries like India, Indonesia, the Philippines and Kenya. The increasingly higher consumption level in the later ages in the rich countries occurs due to health care costs and higher standard of living. The relatively low and downward sloping consumption curve on the poor countries reflects the lower ability to pay, mainly due to informal nature of work and absence of pension, lower accumulation of assets and negligible amount of support from the government. However, as health transition continues apace, the expenditure during old ages is likely to increase for the poor countries as well.

Thus, we must plan for a slowly ageing population where healthcare expenditure will inevitably rise with growing burden of non-communicable diseases (NCD) like cancer, diabetes and heart conditions that require expensive laboratory testing, continued care and monitoring and frequent hospitalisations. While this has been recognized in the policy discourse (Planning Commission 2011), little attention has been directed at the fact that source of income and nature of financial support for the elderly are also changing. Below we consider the importance of the three pillars of support for the elderly—own work and savings, help from extended family and support from the government. Looking at the trends between 2004–05 and 2011–12, we find that each of these three pillars is undergoing substantial changes and is likely to affect the well-being of the coming generations of older Indians.

Data

We document changes in elderly support structure using two rounds of the India Human Development Survey (IHDS), conducted by the National Council of Applied Economic Research (NCAER), New Delhi, and the University of Maryland. The IHDS is a nationally representative household survey. The first round of the survey was conducted during 2004–05 and the second, during 2011–12. IHDS is a multitopic panel survey enumerating more than 200,000 individuals from about 42,000 households. These households spread across 33 states and union territories, 384 districts, 1503 villages and 971 urban blocks, located in 276 towns and cities. Villages and urban blocks (comprising of 150–200 households) formed the primary sampling unit (PSU) from which the households were selected. Urban and rural PSUs were selected using a different design. Urban blocks were selected based on probability proportional to size method. Once the numbers of blocks for each urban area were determined, the enumeration blocks were selected randomly with help from the Registrar General of India. From these Census Enumeration Blocks of about 150–200 households, a complete household listing was conducted and household samples of 15 households per block were selected.

The rural sample contains about half the households that were interviewed initially by NCAER in 1993–94 in a survey titled Human Development Profile of India (HDPI)-1 and the other half of the samples were drawn from both districts surveyed in HDPI as well as from the districts located in the states and union territories not covered in HDPI. The original HDPI followed a stratified random sampling of 33,230 households across 16 major states, 195 districts, and 1765 villages. In 2004–05 survey, 13,593 households were randomly selected for re-interview, of which 11,153 were original households and 2440 were split households, which were separated from these root households but were still living in the village. The data collected through the survey appear to be of high quality and at par with other national level surveys like the National Sample Surveys and the National Family Health Surveys.

The period covering the two rounds of the survey has been an important phase of the Indian economy since during this period various flagship programmes like the National Rural Employment Guarantee Act, National Rural Health Mission, Janani Suraksha Yojana and the Right to Education Act were initiated by the central government and many other programmes like National Old Age Pension Scheme were expanded substantially.

The IHDS collects information on various factors related to the well-being of the older population such as living arrangements, health conditions and health care expenditure, participation in productive work (whether paid or unpaid), and their participation in government-sponsored pension programmes, etc. In addition, IHDS is unique in that it allows us to capture the intergenerational transfers of income and assets among household members.

Socio-demographic Profile of the Elderly Population in India

The second round of the survey has a sample of 21,922 elderly (aged 60 or above) out of which around 68% are from rural areas and 32% from urban areas (Appendix Table A.1). The recent round has around 4000 more elderly than in the previous round. However, due to the ageing of the original sample, the IHDS sample contains a slightly higher proportion of elderly than the national census. For example, the percentage of elderly recorded in IHDS 2011–12 was 10.8%, which is about 2.2% points higher than the recent Census 2011 figures. The proportion of the oldest-old population (aged 80 or above) is also growing fast. Twelve per cent of the elderly belonged to this group during 2011–12 survey, 2% higher than the corresponding figure in 2004–05. Thirty seven per cent of the elderly were widow and 57% were illiterate. While most of the available literature terms elderly as economically dependent, one-third of the Indian elderly continues to work beyond age 60 years.

Economic Security of Elderly in India

Unlike developed countries, with the exception of formal sector workers, few elderly see their later years as a time of retirement that allows them time to enjoy leisure with family, friends and relatives. Given the nature of work, elderly people do during the early years of life, and the lack of savings mechanism and pension system, a majority of elderly in India continues work beyond the legal retirement age for formal sector workers. When they are unable to work, they are usually dependent on the familial transfers in the absence of a sound government support system.

In any society, the economic security of the older population is based on three main sources: (1) own income and savings; (2) support from the extended family, particularly children; and (3) support from the state. Below we examine trends in each of these.

Work and Income as a Source of Old Age Income Security

When we think of the sources of economic support for the elderly, it is important to recognize that own income and assets constitute the largest proportion of income security for the elderly (Fig. 2.3). A large proportion of the elderly households depends on income comes from salaried or wage work; about 87% of the households receive any income from agricultural/business/wage work. Close to 29% of the elderly lives in households which receive income from pensions, interest income and renting property. This also includes rent or crops received for leasing out farmland. These numbers are not surprising since only formal sector workers have access to some form of pension. A large number of workers who are self-employed in family businesses, farming or in manual labour at daily wages do not have this benefit, and their continued employment is mainly driven by lack of alternative income in the form of pensions. Income from property and pension are highest among the better-off segment of population namely educated, affluent income class, high caste Hindu/other religion and those reside in urban areas.

A significant proportion of the older population in India was economically active. These include those involved in salaried or business work, cultivation, farm and non-farm labour and animal care. Nevertheless, the nature of work done by the elderly varies by age, sex and place of residence. About 61% of the older male population in rural areas continues to engage in some economically productive

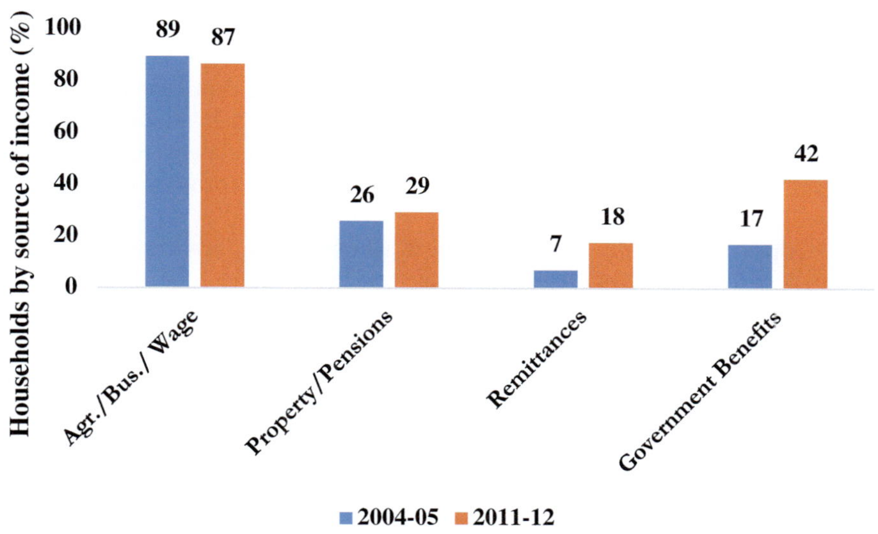

Fig. 2.3 Proportion of elderly households by source of income 2004–12. *Source* Authors' calculations based on IHDS 2004–05 and 2011–12

Table 2.1 Labour force participation rate and type of activities (in per cent) among men and women ages 60 and above in India, 2004–12

Kind of employment	2004–05				2011–12			
	Rural male	Rural female	Urban male	Urban female	Rural male	Rural female	Urban male	Urban female
Salaried	3.00	0.70	8.60	2.30	2.80	0.80	8.90	2.90
Business	6.60	1.50	14.20	1.80	6.80	2.30	15.10	2.80
Cultivation	31.40	10.40	3.40	0.80	36.40	13.10	3.90	0.90
Farm labour	11.50	5.80	3.10	1.10	12.30	7.30	2.60	1.10
Non-farm labour	5.80	1.20	6.80	1.90	14.50	5.60	8.30	2.60
Animal care	35.30	20.80	4.70	3.10	27.30	15.50	4.30	3.70
Doing any work	60.70	31.20	36.90	10.10	60.90	31.10	39.30	12.50

Source Authors' calculations based on IHDS 2004–05 and 2011–12

work compared to 31% of their female counterpart (Table 2.1). In contrast, the elderly in urban areas are less likely to work in their later life. Low rate of labour force participation in urban areas is not surprising since women's work participation rate (for all age groups) in urban areas is lower than in the rural areas. Women (25%) are less likely to work than men (55%) in old age.

Labour force participation among elderly men and women declines with age as one can expect. The rural elderly were mostly engaged in cultivation and farm-related activities including animal care. On the other hand, most of the urban elderly were engaged in family business and salaried activities. A significant share of the elderly in both rural and urban areas was engaged in non-farm activities. In rural areas, salaried work among elderly men and women is almost negligible. Formal sector work imposes much greater entry and exit regulations than work that is family based, such as work on family farms and in family businesses (Desai et al. 2010).

An interesting finding is that there is a substantial increase in non-farm employment in rural areas for both men and women between 2004–05 and 2011–12, which could be attributed to the Mahatma Gandhi National Rural Employment Guarantee Act (MGNREGA). It is possible that many households choose the older members to work in MGNREGA while younger members work in more labour-intensive work (Desai et al. 2015).

Family and Remittances

Family is considered to be the ultimate source of economic, social and emotional support during old age in a society. With the exception of a few areas like Meghalaya and parts of Karnataka and Kerala, India remains a mostly patrilineal society, in which sons are expected to take care of their parents. This reciprocal obligation is bolstered by the fact that in spite of legal reforms, sons continue to inherit parental property, particularly the agricultural land and family residence and in return continue to live with their parents. This tradition is also often held responsible for the strong son preference in India. However, with the gradual decline of employment in the traditional sectors (like agriculture), opening up of new jobs in global settings, and growing individualism, it is often argued that the multigenerational family system is under increasing stress. IHDS finds some support for these expectations and reports five percentage points decline in the elderly co-residence during the last 7 years.

Nevertheless, most elderly persons (72%) continue to live with their children and other family members (Table 2.2). Only 18% elderly live either alone or with their spouse, and the remainder live with children or other family members. The younger elderly often take care of their single and dependent children. But at a later stage in life, co-residence involves living with married sons and their wives. Sixty six percent of the elderly men and 77% of the elderly women live with married children or other members of the family. Living in joint family during old ages is quite common among the richer households and in households with nominal education. Again, elderly in the higher age groups are more likely to be in the joint family than the young elderly, perhaps due to decline in health.

However, with rising prosperity and growing aspirations among both the new and the old generations, the traditional family co-residence may end up as nuclear living arrangements. Knodel and Chayovan (1997) noted only a modest trend towards a reduced co-residence between elderly parents and their children in South

Table 2.2 Living arrangements (in per cent) of elderly aged 60 and above in major Indian states, 2004–12

States	2004–05				2011–12			
	Single	Couple	With unmarried children	With married children/others	Single	Couple	With unmarried children	With married children/others
Himachal Pradesh	2.00	8.83	9.77	79.39	4.89	13.76	6.49	74.86
Punjab	0.71	7.22	9.13	82.93	2.73	11.47	9.90	75.90
Haryana	0.50	9.24	6.98	83.29	1.01	10.93	8.71	79.35
Uttar Pradesh	1.32	9.15	9.26	80.27	2.79	9.60	9.93	77.68
Bihar	0.72	8.25	9.52	81.51	2.16	13.56	10.06	74.22
Rajasthan	3.08	9.36	6.44	81.12	3.91	11.76	8.47	75.87
Chhattisgarh	4.11	11.17	8.80	75.92	5.76	16.19	5.84	72.22
Madhya Pradesh	3.35	9.57	8.38	78.70	5.14	14.69	8.66	71.52
West Bengal	2.06	7.94	17.39	72.61	4.70	14.04	13.59	67.67
Orissa	2.26	10.82	9.03	77.89	4.06	10.85	10.23	74.86
Gujarat	4.74	11.27	8.32	75.67	6.01	11.66	8.50	73.82
Maharashtra, Goa	1.96	9.09	7.29	81.66	3.14	14.12	11.65	71.10
Andhra Pradesh	3.54	16.07	7.54	72.86	6.88	21.67	7.45	64.00
Karnataka	2.18	5.96	13.39	78.46	2.62	9.45	10.82	77.11
Kerala	2.70	9.11	10.93	77.25	2.07	14.33	6.99	76.61
Tamil Nadu	4.82	16.91	17.00	61.27	8.19	22.06	14.02	55.73
India	2.44	9.81	10.79	76.95	3.98	13.58	10.7	71.75

Source Authors' calculations based on IHDS 2004–05 and 2011–12. States with elderly sample 500 or above have been listed here

Asian Countries. On examining the perception of the aged themselves on their status in the family and community in the wake of changing societal structure in India, Rajan and Kumar (2003) found that the aged feel that they have lost status as a result of old age or their retirement from work. Diwan et al. (2011) observed a reduced filial obligation expectation among English-speaking older Asian Indian immigrants in Atlanta, GA. They found that very few respondents preferred to move with their children, but they liked to move closer to children after retirement. So, it becomes pertinent to examine whether similar expectations prevailed among native Indians. We cannot examine the expectation since we do not have data for the same. But, change in living arrangements among the elderly between IHDS 1 and 2 can provide us some insight on the issue.

While there remains some difference in change in magnitude, all the major states in India face a stark rise in families, where elderly either lives alone or with spouse only (Table 2.2). Though most elderly still live with their children and other family members, a significant decline has been recorded in families where the elderly lives with married children and other family members. Less than 1% of the elderly in Haryana live alone, in contrast, more than 8% of the elderly in Tamil Nadu live alone. We do not see any unidirectional trend in the living arrangements among the elderly living with unmarried children. This gives a sense of consistent weakening of family co-residence in Indian states as well.

It is important to note that men and women face ageing differently. The recent round of the data shows that among elderly while 57% women were widows, only 17% men were widowers. The longer life expectancy of women, on an average, and the normative age gap between husband and wife make widowhood more likely for women than for men. This also accounts for the fact that more women live with their married children than men.

Declining co-residence (about 6% percentage point between 2004–05 and 2011–12) does appear to be a harbinger of doom for the Indian extended family. Children continue to provide financial support for their parents even when they do not co-reside. About 18% of the elderly households received any income from remittances in 2011–12, up by 7% than in 2004–05. Remittances seem more likely for couples or single individuals whose children live elsewhere. When the elderly live with family members, they receive help via co-residence and do not receive monetary transfers (Desai et al. 2010).

Nonetheless, remittances do not replace co-residence in ensuring incomes of the elderly households. We find that though the proportion of households receive any income from remittances has gone up significantly during the last 7 years (Fig. 2.3), the amount received from remittances has remained more or less the same (Rs. 40,651 in 2011–12 compared to Rs. 36,332 in 2004–05) (Table 2.3). At the population level, the total amount received from remittances is Rs. 7158.

Table 2.3 Sources of income of the households with elderly member, 2004–12

	Per cent households with any income from				Average annual income (in Rs.)							
					In households with elderly				In whole population			
	Agr./ Bus./ Wage	Property pensions	Remittances	Government benefits	Agr./ Bus./ Wage	Property pensions	Remittances	Government benefits	Agr./ Bus./ Wage	Property pensions	Remittances	Government benefits
	2004–05											
All India	89.4	25.9	10.7	17.1	93638	49440	36332	3267	83400	12138	3904	559
Gender												
Male	89.8	27.4	10.4	14.7	93734	52365	36417	3316	83921	13699	3769	489
Female	88.9	24.3	11.1	19.5	93539	46059	36251	3229	82867	10541	4041	631
Age												
60–69	90.4	25.5	10.6	15.8	91104	51256	34026	3186	82134	12467	3612	503
70–79	88.2	25.1	10.9	19.3	97482	49032	36884	3186	85704	11556	4011	614
80+	86	30.4	11.2	19.2	99186	40980	48326	3900	84850	11720	5404	749
Family type												
Single	44.8	35	21.6	29.6	16311	18273	17296	3326	7304	6240	3737	984
Couple	57.6	40.4	14.6	15.4	55245	58394	28666	4116	31605	23391	4188	634
With Unmarried Children	90.1	27.1	6.7	9.5	70992	54610	33533	3285	63699	14560	2253	313
Joint	94.7	23.6	10.5	18	100802	48120	39196	3170	95187	10551	4104	570
	2011–12											
All India	86.5	29.3	17.6	42.3	115962	67379	40651	4965	99731	19745	7158	2102
Gender												
Male	87.1	30.2	18	38.3	119884	74001	41179	5284	103759	22383	7417	2022
Female	85.9	28.4	17.2	46.2	112215	60736	40131	4716	95936	17259	6914	2177
Age												
60–69	88.4	28	17.6	39.7	112489	69097	40864	5060	98814	19354	7196	2010

(continued)

Table 2.3 (continued)

	Per cent households with any income from				Average annual income (in Rs.)							
					In households with elderly				In whole population			
	Agr./ Bus./ Wage	Property pensions	Remittances	Government benefits	Agr./ Bus./ Wage	Property pensions	Remittances	Government benefits	Agr./ Bus./ Wage	Property pensions	Remittances	Government benefits
70–79	83.6	32	17.4	45.9	119707	66288	38413	4896	99517	21206	6689	2246
80+	83.7	29.5	18.1	47.3	125841	61827	44771	4714	104965	18240	8093	2227
Family type												
Single	44.3	29.7	27.6	51.1	21179	35761	13897	4071	9332	10613	3830	2081
Couple	61.2	35	21.6	34.3	47076	80293	31572	5847	28293	28111	6812	2005
With Unmarried Children	89.6	26.9	12.4	31.6	107538	81017	30696	5505	96023	21824	3793	1741
Joint	93.2	28.6	17.1	45	128226	64284	46286	4837	118814	18357	7910	2175

Source Authors' calculations based on IHDS 2004–05 and 2011–12

Public Transfer Scheme

Though evidence suggests that work and family are the major supports for the elderly, the government-sponsored safety net programmes also play a small role in it. India has a social safety net programme, National Social Assistance Scheme (NSAP) that has functioned since 1995 and the National Old Age Pension Scheme (NOAPS) is an integral part of it since the inception. Under the NOAPS, a central contribution of Rs. 75/month was given to the destitute elderly aged 65 years or above. States supplement this, although the amount varies between states. Over time, the coverage, the criterion of destitution and pension amount, has changed with the change in its jargon. The Indira Gandhi National Old Age Pension Scheme (IGNOAPS) provides support to the elderly of age 60 years or above who belong to a BPL household. The pension amount by the centre has increased to Rs. 200/month up to age 79 years and Rs. 500/month to the oldest-old (age 80 years or over). Eligible elderly also receive 10 kilograms of food grains per month with no cost through Annapurna Yojana.

The role of public transfers in elderly households has grown sharply in recent years. The recent round of the IHDS shows that 42% elderly households receives any government benefits such as old age pension, widowhood pension, disability pension up from 17% in 2004–05 (Fig. 2.3). Increasing proportion of elderly receiving income from IGNOAPS—7% versus 18%—plays an important role. In recent years, eligibility for IGNOAPS has changed from 65 years to 60 years and greater efforts have been made to enrol eligible beneficiaries. Most government schemes appear to be targeted towards the vulnerable sections of the society; rural residents, females, Dalits and Adivasis are more likely to receive government assistance. More than half of the elderly living alone receive government pension. Close to half of the elderly living in multigenerational joint families are also getting this benefit. However, payments through these schemes are only modest in size. For households that receive transfers, annual benefit is Rs. 4965 on average; for population as a whole, it is Rs. 2102 per year.

Caring for the Elderly in a Changing World

Over the last two decades, several states of India have achieved total fertility rates below 2.1 (known as below replacement fertility). Even the most populous states like Uttar Pradesh and Bihar are also showing a rapid decline in fertility rate, which eventually leads the Indian society towards its ageing phase. However, declining fertility is also associated with declining co-residence and declining proportion of working age adults per elderly household. In IHDS data, average number of working adults of ages 15–59 has declined from 2.9 in 2004–05 to 2.5 in 2011–12. This decline in workers per household coupled with declining co-residence suggests a need to rethink our reliance on extended family for economic and social support of the elderly.

Each of the three pillars of economic support in the old age—work and personal savings, extended family and state—must be prepared for increasing demands in the decades to come. Below we outline some of the issues that deserve consideration.

(1) Work and Savings: With some exceptions, retirement ages around the world have been rising (Lee and Mason 2011). Even in India, retirement age in some sectors such as universities has risen. We may want to think about increasing retirement ages in formal sector jobs. Economic growth and labour shortages in some sectors provide us with an opportunity to do so. More importantly, we may also need to create vehicles for increased private savings and investment. A variety of schemes are currently available to the middle classes but safe and accessible methods of savings are often not available to the poor who must deposit small amounts and are often not financially sophisticated. Current discussion on financial inclusion should be expanded to include micro-savings. Improving financial inclusion will also aid the elderly households in receiving remittances from family members living elsewhere, a growing source of income as documented above.
Moreover, present inheritance laws and laws governing agricultural property make it difficult for the elderly to sell or borrow against property. Simplification of these laws and regulations and reduction of black money component from real estate transaction may make it easier for the older population to self-finance their old age.
(2) Learning to Live with Declining Familial Support: With declining fertility and increasing migration, the decline in extended families observed between 2004–05 and 2011–12 is simply the tip of the iceberg. We should expect more and more elderly individuals to live alone or with their spouses. In addition to day-to-day care, it may also affect their ability to obtain health care. Thus, as we develop health systems for diagnosis and treatment of ailments of older ages, particularly for non-communicable diseases like diabetes and heart diseases, we may need to think about the fact that many elderly patients will not have family members available to help them in negotiating access to health care.
(3) Developing Infrastructure for Public Transfers: While it seems unfeasible that India can provide substantial income support to all elderly via a social pension programme, we can draw substantial lessons from the experience of other countries. Few countries have set up social security programmes overnight. Development of social pension programmes requires both building political support for these programmes and developing administrative capacity. Thus, it may make sense to start with a universal programme of modest pensions. Even now, a large number of elderly households receive transfer payments of some kind. For example, the IHDS data suggest that 42% of the elderly households receive modest transfers. Instead of a complex web of schemes such as Annapurna Yojana, IGNOAPS and income tax benefit schemes, it may make

sense to consider a universal modest social pension programme that benefits all elderly. This will form the backbone of a social security programme as income of the country grows and we seek to provide greater income security for the elderly. However, developing efficient transfer mechanisms via Direct Benefits Transfer or some other distribution channel will be essential in order to reduce corruption and leakage.

Acknowledgements India Human Development Survey was funded by grants *R01HD041455* and *R01HD061048* from the US National Institutes of Health and a supplementary grant from the Ford Foundation. Data analysis was the funded by the UK government as part of its *Knowledge Partnership Programme* (KPP). We gratefully acknowledge inputs from Prof. Reeve Vanneman and Prof. Amaresh Dubey and participants in a roundtable on 'Caring for the Elderly in India: Challenges for a Society in Transition' at NCAER (New Delhi).

Note: This is a slightly revised version of our paper published in Economic and Political Weekly Barik, D., T. Agrawal, and S. Desai. 2015. 'After the Dividend: Caring for a Greying India'. *Economic and Political Weekly* 50(24):108.

Appendix

See Table A.1.

Table A.1 Socio-demographic profile of the elderly population in India, 2004–12

	2004–05	2011–12
Sex		
Male	50.1	48.0
Female	49.9	52.0
Age		
60–64 years	35.2	33.1
65–69 year	25.8	25.9
70–74 year	19.5	18.5
75–79 year	9.6	10.8
80 years and above	10.0	11.8
Marital status		
Unmarried	0.6	1.0
Married	63.1	61.9
Widowed	35.8	36.8
Divorced/Separated	0.4	0.3
Educational level		
Illiterate	59.6	57.41
Up to 5th Std.	17.5	17.84

(continued)

Table A.1 (continued)

	2004–05	2011–12
Secondary level	15.3	17.43
Metric but non-graduate	2.3	3.13
Graduate and above	3.1	3.99
Missing Education Data	2.2	0.2
Religion/Caste		
Forward caste	26.4	25.8
OBC	35.3	35.3
Dalit	17.7	18.4
Adivasi	6.3	6.7
Muslim	9.7	9.9
Christian, Sikh, Jain and Others	4.7	3.9
Place of residence		
Rural	70.6	67.9
Urban	29.4	32.1
Working status		
No	60.4	67.3
Yes	39.7	32.7
Total	17904	21922

Source Authors' calculations based on IHDS 2004–05 and 2011–12

References

Census of India. (2011). *Census Provisional Population Totals*. The Registrar General & Census Commissioner, India. Retrieved February 14, 2013.

Desai, S. B., Dubey, A., Joshi, B. L., Sen, M., Shariff, A., & Vanneman, R. (2010). *Human development in India: Challenges for a society in transition*. New Delhi: Oxford University.

Desai, S., Vashishtha, P., & Joshi, O. (2015). *Mahatma Gandhi national rural employment guarantee act: A catalyst for rural transformation*. New Delhi: National Council of Applied Economic Research.

Diwan, S., Lee, S. E., & Sen, S. (2011). Expectations of filial obligation and their impact on preferences for future living arrangements of middle-aged and older Asian Indian immigrants. *Journal of Cross-Cultural Gerontology, 26*(1), 55–69.

Knodel, J., & Chayovan, N. (1997). Family support and living arrangements of Thai elderly. *Asia-Pacific Population Journal, 12*(4), 51–68.

Lee, R., & Mason, A. (2011). Lifecycles, support systems, and generational flows: Patterns and change. In R. Lee & A. Mason (Eds.), *Population aging and the generational economy: A global perspective*. Cheltenham, UK: Edward Elgar.

Planning Commission. (2011). *High Level Expert Group Report on Universal Health Coverage for India*. http://planningcommission.nic.in/reports/genrep/rep_uhc0812.pdf.

Rajan, S. I., & Kumar, S. (2003). Living arrangements among Indian elderly: New evidence from national family health survey. In *Economic and Political Weekly* (pp. 75–80).

Chapter 3
'Fragile Body and Failing Memory': The Construction of Care for the Elderly by the Laws and Policies in India

Deblina Dey

Abstract The genesis of legal discourse around care for the elderly in India can be traced to the international gerontological discourse, which pathologizes ageing and attempts to rectify it by providing universal, technical solutions to the elderly. It completely ignores the subjective articulation of needs by elderly in different socio-cultural locations. Critically analysing relevant judgements and policies, the chapter examines how the elderly are named in policy and law, and how their needs are interpreted. The study finds that the State re-enforces the dependency of the elderly through reiterative nomenclatures, which locates empowerment not solely in terms of their rights as senior citizens, but in terms of their assumed physical fragility. Also, the legal discourse around care in India valorizes the notion of care for the elderly as a matter of the family. The study highlights that the juridical notion of care is limited to 'maintenance' of elderly parents, in which the right to care is limited to food, clothing, residence and medical treatment of elders by their potential heirs. Such juridical notion of care bypasses needs of the elders who are childless or are single women. Care remains a uni-directional flow of material things from a person (heir) with greater resources to the elderly who seen as incapable of doing so. Even though this is respite to many, yet juridical care does not address the symbiotic and dialogic nature of care, as is in the case of Chinese law.

Keywords Care · Law · Medicalization · Vulnerability · Needs

Deblina Dey (✉)
Jindal Global Law School, O.P. Jindal Global University,
Sonipat 131001, India
e-mail: ddey@jgu.edu.in

Deblina Dey
Centre for the Study of Law and Governance, Jawaharlal Nehru University,
New Delhi 110067, India

S. Irudaya Rajan and G. Balagopal (eds.), *Elderly Care in India*,
DOI 10.1007/978-981-10-3439-8_3

Introduction

Care for the elderly in India is traditionally associated with familial care or care within the domestic sphere. This notion of care is valorized in the legal and policy discourse in India. Taking recourse to extra-familial forms of care services is generally stigmatized and is indicative of the 'Bad Family' with the evil and 'modern' daughter-in-law who fails to take care of the elderly in-laws (Cohen 1998). Within the State's discourse, as evident through policies and judgements, westernization and urban ways of life (and its associated attitudinal changes) are depicted as agents of corrosion of the Indian joint family system that is depicted as the ideal location for ageing members. This study explicates the State's conceptualization of care or the 'juridicalization'.[1] Thus, in the current context, bringing care into the ambit of juridical practice connotes its juridicalization of care for the elderly through the legal provisions as well as policy enactments.[2] It is also important to note that the State's discourse on care for elderly or senior citizens in India is influenced by the international gerontological discourse. The study traces the trajectory of discourse around old age as originated in the West and how it influenced the way ageing was perceived in India. Critically analysing relevant laws and policies, I examine how the elderly are named in policy and law. I argue that although the State upholds the figure of the senior citizen as a right-bearing individual, it also re-enforces the dependency and vulnerability of the elderly as a special category of persons. This study is an attempt to uncover the legal mechanics behind the *performativity*[3] of an ageing identity. I conclude by showing that discourse on empowerment of elderly is structured around the identification of elderly as a weak individual, reinforcing the idea that care is for the 'needy' persons, somewhat positioned inferior to oneself in terms of various resources.

Emergence of the Discipline of Ageing and Impacts on Policy Discourse: A Historical Overview

Studies on ageing gained pace with two historical developments around the nineteenth-century Europe. Firstly, there was the rise of systematic census data collection as expressed in statistics on life expectancy at various stages of life

[1]Black's Law Dictionary (1st edition), [also available online at http://blacks.worldfreemansociety.org/1/J/j-0663.jpg] defines 'juridical' as 'relating to administration of justice' and 'done in conformity to the laws of the country and the practice which is there observed' (1891:663).

[2]Law and policy are mentioned together in several places in the study because they both have one common ground in expressing the State's views of care—one with sanctions and the other without any, the latter being a framework but not rule-bound. However, it does nowhere suggest a convergence between the two.

[3]See Butler (2007).

(Green 1993:42). Secondly, medical practice underwent a significant change in which attention was now given to treatment of disease in individual persons (Foucault 2003). The 'medical gaze', which Foucault (2003) highlighted to be prevalent in medical practice around the nineteenth century, made possible the detailed diagnosis of bodies in general. These two developments according to Green (1993) were fundamental to understanding ageing and the emergence of the discipline of gerontology. From a 'generalized notion of senescence as depleted vitality', ageing was to be found in specific locations of organs in old bodies, which would require specific kind of treatment (Green 1993:43). This highlights the change in the discourse about ageing from the 18th to 19th centuries and the subsequent emergence of the elderly as a medicalized object of study (Green 1993).

The initial stages of gerontological imagination were predominated by the biomedical model of ageing, much in tune with the idea of the 'medical gaze' and rootedness of the problem of old age in the body of the person (Katz 1996). At the first World Assembly on Ageing, held in 1982 at Vienna, the Plan of Action charted by the United Nations was accepted by all without any disagreements. Based on the observation of a swelling ageing population globally, the countries recognized the need to address the concerns and 'needs' of the elderly through developing the capacities of countries to deal with their needs and recognizing the requirement for 'international technical cooperation' among developing countries (Bose and Shankardass 2004:155). Subsequently, in 1999, the Indian government initiated a series of national-level policies with the National Policy for Older Persons (henceforth referred to as NPOP) as a foundational policy for the elderly in India. Cohen (1998) argues that until the World Assembly on Ageing in 1982 in Vienna, there was no concept of the 'senior citizen' in India; as Cohen (1998) would argue, it was as if there was 'no ageing in India' evidenced by the near absence or inadequacy of social provisions for the elderly. Thus, one finds the discourse on care completely missing in the public sphere in India prior to 1999 Policy. Analysing the discourse generated at the policy level enables one to discern the ideological basis of treatment towards elderly, which in due course of time becomes institutionalized. The disciplinary discourse around gerontology that is relatively new in India as compared to American and European countries also must be subject to scrutiny in order to understand the ways in which it reinforces, recreates and institutionalizes certain modes of behaviour towards the elderly. The adoption of the 'old age' as a governmental category came about only after the First World Assembly in 1982. Prior to that elderly were not seen as a unique, separate category and were mentioned along with other categories. For instance, the Constitution of India in its Directive Principles of State Policies under Article 41 stated:

> The State shall, within the limits of its economic capacity and development, make effective provision for securing the right to work, to education and to public assistance in cases of unemployment, old age, sickness and disablement, and in other cases of undeserved want.

Clubbing together the elderly with the other categories as mentioned above, the term 'undeserved want' seems to reinforce a certain stereotype against the elderly as a baggage to the State exchequer. In initial constituent assembly debates, with reference to one in 1948, a certain Shri Syamanandan Sahaya, who was the Bihar General at that time appealed that this phrase be replaced by the term 'deserving relief', but such change in terminology was not seen important. It is not wrong to argue that the stigma associated with some category such as 'old' does not find escape and can become the basis of neglect and discrimination.

An International Plan of Action, which had an underlying assumption of a 'single ideology of gerontological practice', was charted out for all the countries at Vienna in 1982 (Cohen 1998:94). The outcome of the Vienna meet suggested an international gerontology that reflected the evolutionist idea of a universal world-view. 'Old age' was conceptualized as a typical type of social problem which could be dealt through particular remedial measures like providing services like day care centres, laundry services, home delivery of meals and the like (Cohen 1998). Such an approach failed to take into account the heterogeneous category of the elderly, their unique needs and cultural variations in understanding ageing. Many policies of contemporary times in India follow the logic of solving old age issues by providing facilities to them. It will require another study to look at the implementation aspect.

Critical Analysis of a Few Salient Policies

The National Policy for Older Persons (1999) set in motion a discourse around the greying of population in India and aimed to put in place the necessary provision for this section of population. This policy was conjured in the Ninth Five Year plan and identified the elderly as not just beneficiaries of the services but highlighted the need for their 'active involvement in developmental activities' (Liebig and Rajan 2005:222). The NPOP stressed the productive capacities of the elderly for the first time. The years of the First World Assembly 1982 may be considered as a watershed moment in the discourse of gerontology, as it was the first instance of a worldwide recognition of an increasing elderly population globally. Similarly, 1999 was the watershed moment in the discourse of gerontology in India, because prior to this, there were no national-level policies which took into account the socio-cultural changes in the family and work sphere. Only legal procedures regarding property transfer and pension schemes existed in prior to 1999. Health, shelter and recreation education all of this were for the first time brought under the rubric of a national policy, viz. NPOP (1999).[4]

Demographically, NPOP (1999) signals out an alarming condition and acknowledges that ageing which is a global phenomenon has 'hit Indian shores as well'. It is interesting to note the language used in this policy. It spells out a sense of

---

[4]Other significant laws and policies which were enacted in the years following the NPOP were the Maintenance and Welfare of Parents and Senior Citizens Act (2007), Indira Gandhi National Old Age Pension Scheme (2007), Integrated Programme for Older Persons (2008), National Policy for Senior Citizens (2011). Of these, only a few, ones which are more foundational, will be discussed in detail.

foreboding that people now are living longer due to improved life expectancy. The NPOP (1999) points out that middle-and upper-income groups will be economically secured relatively and are expected to lead 'active, creative and satisfying life'. It recognizes the Western impacts of urbanization, changes in lifestyle and work, family system increasingly becoming nuclear, women choosing to work out and thereby 'desire not to be encumbered by caring responsibilities of old people for long periods' thus reducing the caregiving time for the elderly at home. The NPOP identifies the above factors responsible for the corrosion of traditional ideas of *seva* (care) of the elderly by family members. The policy aims at restoring their positions as right-bearing citizens in society and hence prevents their identity crisis vis-à-vis other age group citizens (NPOP 1999). The policy holds that the older persons are a resource and thus should be empowered, given affirmative rights, tax concessions, proper health care to enable them to live the 'last phase of their life with purpose, dignity and peace' (NPOP 1999:3).

By clearly stating what is valued as the above, there is clear message about what is devalued and seen almost as an evil. We know that 'any norm is constituted through a citation of its exceptions' and therefore, analysis of the exceptions becomes important here (Boucher 2006:116). The ideal form of *seva* and care as conceptualized by law is the care given by family members. In the 'welfare' section of the NPOP (1999), it is specified in the policy that institutional care would be the 'last resort when personal circumstances are such that stay in old age homes becomes absolutely necessary'. This clearly depicts that living in old age homes comes with a stigma and is less desirable to elderly people. The NPOP (1999) mentions that whatever forms of non-institutional support services are to be provided by the State it would be on the basis of 'sharing of family's caring responsibilities' showing the tragic inevitability of social transformations towards lack of caregivers for the elderly within the increasingly smaller families.[5] Also, the policy values an 'age-integrated society' (NPOP 1999:19) highlighting the importance of intergenerational connectedness, the absence of which automatically qualifies as pathological. The policy stretches to the extreme that in order to grapple with corrosive forces of Western culture, the children of aged parents will be provided monetary benefits in order to 'encourage' them to co-reside with their parents (NPOP 1999:10). It seems as though there is an alliance between the State and family, as suggested by Donzelot (1997) eliding the fact that co-residence does not ensure dignity, respect and autonomy for the elderly. The policy assumes that there is a need for group living at this stage of life (NPOP 1999:8). It could be pointed out

[5]The existing laws not only reinforce the stigma associated with residing in old age homes but shrink from its responsibilities in providing basic care facilities to elder persons, by holding the family responsible for the condition of elderly members. This situation overshadows the alternative ways of living among a host of elderly persons. For instance, in contemporary society, with the care market eulogizing and glamorizing old age itself by propagating the idea of 'ageing as luxury', one finds that old age homes in many instances are not the case of last resort but a willing and planned relocation for some, particularly those who can afford the different ranges of. See http://www.businesstoday.in/moneytoday/real-estate/old-age,-new-home/story/6258.html.

that group living does not necessarily mean that the loneliness experienced by the elderly is mitigated, though in many cases it could. Also, by assuming or generalizing the need for group living at 'this stage of life', it enables the caregivers to secure a greater control over the lives of the institutions' residents by forcing them to live and work together. The problematic aspect of any policy is that is takes the needs of the beneficiaries as 'pre-given' and the policies redefine them according to 'system-conforming satisfactions' (Fraser 1989:156). This is visible in the aforementioned instance. It can be seen that even though the ageing elderly are considered right-bearing citizens, they are clearly understood as 'passive citizens' (Fraser 1989:156).

However, within the same policy, one could discern the language of care anchored on the idea of 'empowerment' of the elderly (NPOP 1999). The policy claims this enables them to

> …acquire better control over their lives and participate in decision making on matters which affect them as well as on other issues as equal partners in the development process (NPOP 1999:3).

This tone of administering care also reflected in the Madrid Meet in 2002, the Second World Assembly 20 years after the first meeting at Vienna. The central theme of this meeting was 'building a society for all ages'. The Madrid Action Plan revealed that concern of an increasing aged population needs to be located vis-à-vis the Millennium Development Goals (MDGs). The aged population would become a 'problem' for developing countries because such countries 'became old before they become rich' (Bose and Shankardass 2004:175). This is in sharp contrast to, according to the Director General of World Health Organization Dr. Gro Harlun Brundtland, developed countries who become old after they become rich. In other words, the developed countries have the capacity to deal with an increasing ageing population, whereas developing countries do not. It was a matter of concern that the more number of years to live meant greater number of years in poverty for the elderly in the developing nations. Thus, the Political Declaration that was adopted at the Second World Meeting in 2002 was the juxtaposition of development goals of the countries with ageing population and an attempt to eradicate poverty.

It can be argued that with the influence of international gerontology, there emerged a medicalization[6] of ageing. Health interventions through various policies localize the pain or illness in specific bodily parts and States, which have universal diagnostic criteria. However, it does not make sense to discount local narratives of pain, suffering, health and body which might vary according to class and regions. One must understand that defining old age as a phase having universal pathological conditions is problematic. The techno-centric, service-oriented model propagated by the international gerontological discourse at the First World Assembly in 1982 in Vienna sidelined many of the cultural variations and leaves no space for subjective evaluations of needs of elderly without sufficient discourse on how the need

[6]Medicalization is a 'process which more and more of everyday life has come under medical dominion, influence and supervision' (Zola 1983:295).

to be dependent is articulated by the elderly themselves or the complex process by which it is recognized by the caregiver. For instance, one may analyze the scheme, National Programme of Health Care for the Elderly (NPHCE 2010), which mentions that non-communicable diseases (lifestyle related and degenerative) are extremely common in older people irrespective of socio-economic status. The scheme would ensure that adequate care is provided to the elderly in the event of such diseases. However, the very diagnosis of disease does not always depend on the elderly individual's biological condition but rather the expression of that need brought about by other persons in her life (Kleinman 1988). The definition of a pathological condition can depend upon the caregivers' own well-being, for instance the level of stress experienced by the family members. Senility is dialogic and attributional (Cohen 1998). Understanding ageing as a phenomenon involves two bodies—the ageing body and another person who would explain the changes happening in that other ageing body.[7]

Let us now turn to the most recent policy, The National Policy for Senior Citizens (NPSC) in 2011, a reworking of the NPOP with only a few additions. In this, significant recognition has been given to women's issues. For instance, it has been recognized that a higher number of women have to deal with loneliness than men. Also, ageing in rural and urban areas have found separate mention in the 2011 Policy. India has been a supporter of the Plan of Action generated at the Madrid Meet in 2002. The Madrid Action Plan 2002 envisages a 'society for all ages', where discrimination would be ideally done away with. The National Policy 2011 also aims to create a 'barrier free and age-friendly society' (NPSC 2011:4). While this may sound well intended, it remains to be known how an age-friendly society could be created. One addition with regard to encouragement that the government plans to give towards enabling post-retirement employment arouses curiosity as to how it could be achieved when unemployment rate for the youth is itself on the rise?[8] As per the NPSC (2011), welfare measures like providing universal identity through Aadhar cards will be made possible for ageing women so that they can avail the different government schemes. However, it can be seen that many women who are destitute do not have a proper address to ensure themselves the Aadhar in the first place. Many of them, according to the NPSC (2011), would perhaps like to work as domestic maids. However, without strict regulation, chances are that they will be exploited. Thus, it can be concluded that even after more than a decade, policy has not materialized effectively into practice (Paltasingh and Tyagi 2012:14). This challenges how even well-intended policies can often be at odds with the needs of the people in reality. Let us now analyse the legal provisions for the elderly.

---

[7]Brijnath (2014) describes the diagnosis of senility and subsequent institutionalization on the basis of complaints and threats by neighbours to the daughter to hospitalize the father since he caused nuisance in the absence of the daughter after her migration to another city.

[8]'Unemployment rate in India is showing an increasing trend since 2011 when it was 3.5%. The same rose to 3.6% in 2012 and climbed to 3.7% last year' (See http://timesofindia.indiatimes.com/business/india-business/Unemployment-levels-rising-in-India-experts-say/articleshow/29403619.cms).

The Right to Care and Creation of 'Subjects of Power and Agents of Resistance': Analysing the Maintenance Act of 2007

Important aspect of legalizing care for the elderly in India is expressed by criminalizing the lack of care. This can be gauged from the Section 125 of Criminal Procedure Code (CrPC) of 1973, which codify care for parents or aged in the terms of 'maintenance' of the elderly. One may note that the juridicalization of care in India does not expand beyond economic concerns of the parents. Thus, it is only the lack of maintenance, in terms of payment of a monthly allowance to the parent that can legally coerce the child to take care of the parent economically. A person 'with sufficient means' cannot neglect his or her parent according to the CrPC, and on breach, the person becomes liable to that fine or imprisonment. An improvement upon this section of the CrPC is the Maintenance and Welfare of Parents and Senior Citizens (henceforth the MW Act of 2007). This is by far the most important piece of legislation for the elderly, which takes into account economic concerns of not only a parent (as was the case with the CrPC 1973) but also senior citizens of the country who are unable to maintain themselves. This Act provides a time frame for disposal of the case within ninety days from filing a petition. It goes beyond the ties of child and parent and includes any relative or such person who is compelled to maintain the senior citizen, if the person is entitled to the property after death of the senior citizen. The CrPC does not deal with case related to fraudulent property transactions.

Another area of concern which demonstrates the urgency of strict legal provisions is that of increasing cases of elder abuse. According to the HelpAge India Report for the World Health Organization (Soneja 2011), elder care has always been a matter within the family and elder abuse was until very recently not legally an important concern. According to their study, neglect though the most common[9] form of abuse is seldom named as 'abuse' by the victim themselves. The elderly respondents mentioned only 'extreme behaviour of violence' when they referred to abuse. In another survey conducted by HelpAge India in 2014,[10] it was found that most of the elder abuse is from their own children particularly the son (maximum abuse), daughter-in-law and daughter. The laws could provide a scope for an expression of resistance to maltreatment of parents by children. The MW Act, 2007, safeguards, albeit limitedly, only some of these concerns. According to this Act, the following rules of application for maintenance hold for:

---

[9]About 48.7% of elderly abuse cases confirm neglect. See http://infochangeindia.org/human-rights/analysis/indias-senior-citizens-finally-get-a-hearing.html.

[10]The survey found verbal abuse (41%), disrespect (33%) and neglect (29%) as the most commonly occurring forms of abuses experienced by the elderly. See http://timesofindia.indiatimes.com/india/Abuse-on-elders-increase-by-50-this-year-Helpage/articleshow/36503113.cms.

> (i) parent or grand-parent, against one or more of his children not being a minor; (ii) a childless senior citizen, against such of his *relative*. Secondly, the obligation of the children or relative, as the case may be, to maintain a senior citizen extends to the needs of such citizen so that senior citizen may lead a *normal life*. Thirdly, the obligation of the children to maintain his or her parent extends to the needs of such parent either father or mother or both, as the case may be, so that such parent may lead a normal life. Fourthly, any person being a relative of a senior citizen and having sufficient means shall maintain such senior citizen provided he is in *possession* of the property of such senior citizen or he would inherit the property of such senior citizen: provided that where more than one relatives are entitled to inherit the property of a senior citizen, the maintenance shall be payable by such relative in the proportion in which they would inherit his property.[11]

The question of inheritance of assets by 'relatives' becomes problematic for instance in the case of childless senior citizens. This is seen in *Lotika Sarkar v. Preeti Dhoundial and Ors* (2009)[12] where a childless widow, and an eminent law professor LS had been fraudulently made to sign a gift deed of her house to the D family who were not seen by her as her legal heirs. According to the case reporting, LS had signed the papers believing that she was leaving the maintenance of the house to the D family, especially Mrs. D, who had been willingly residing with her after LS's husband's death. She by no means consented to the legal hand over of the property to the D family and though that the maintenance of the house would be done through her own savings. However, the signature on the deed in reality confirmed that she has on her own transferred her property at Hauz Khas Enclave in New Delhi to the D family. The proceedings highlighted that she had also been made to sign under grief and shock of her husband's demise coupled with 'loss of memory', 'brainwashing' and 'tutoring' as the judgement reads. Mr. D mentioned that they 'assumed' that they would inherit the property, as she has no one else to look after her. Whether LS would have given the property to any other person or not given it to anyone could be debated. In such cases, it becomes difficult to determine who would 'maintain' such a senior citizen if the law is rooted to giving responsibility of maintenance to those who would inherit property, when the latter might itself be difficult to determine.

The law also sets 'normal life' as a goal while forgetting it can have subjective elements. While the implication of this phrase can be guessed, this word is troubling because normality is context specific and very subjective. There is no provision in the law which pertains to emotional abuse or neglect of elderly parents. In reality, it is difficult to prove emotional abuse but the law should also define 'normal' if it wishes to use the term. Another aspect of concern is that the MW Act also does not address the question of those senior citizens who do not have either children or

---

[11]The above is quoted verbatim from the MW Act enacted in 2007 under chapter II.

[12]This case was taken as a precedent for a number of other cases henceforth, for example, in *Anbalagan v. The Sub Divisional Executive Magistrate and Revenue Divisional Officer* (2012) W. P.No. 12145 of 2012. In this case, the right to residence in own property of the parent is highlighted. The MW Act comes as a rescue to interference by one's own children against retaining one's own property. Similar to LS case, also *see Wp(C).No. 17245 Of 2012 (E) vs By Adv. Sri. Varghese P. Chacko* (http://indiankanoon.org/doc/170672315/).

property. While understandably this law addresses problems faced by those elderly who have children, it needs to be fine-tuned further to include those senior citizens who are not parents. Those who are not parents will have peculiar conditions of living and question of maintenance becomes significant to them. By providing a half-baked law, the State seems to choose to continue to be in this inertia of inaction for growing elderly population in the country who have no caregivers, no possessions or shelter, and cannot avail basic care provisions, to continue a life of dignity.

The MW Act only refers to the material deprivation or abuse faced by the elderly but does not take into account for instance emotional abuse or neglect, which as HelpAge India Report highlights as the most important form of abuse. The Maintenance Tribunal often returns the cases, if the parent appeals not for maintenance but against forms of abuse at the hands of children. 'Care' in legal discourse is circumscribed to food, residence, clothing and medical expenses and treatment.[13]

In this context, one may compare Indian and Chinese legal provisions since both countries have similar cultural values of family care of the elderly members and both are currently grappling with the transformation of family system due to processes of urbanization. A Chinese law enacted in 2013 called the 'Protection of the Rights and Interests of Elderly People', makes it mandatory for the children to visit their parents frequently. Thus, 'A Chinese Virtue is now the Law' as reported in *The New York Times* (Wong 2013). The Chinese law sues the children for not visiting the parent and thereby neglecting the parent. This is a clearest formulation of juridicalizing care. In India, on the other hand, care is understood in economic terms of protection of property and monthly allowance by children. Suing children in terms of fine or imprisonment happens only when there is a negligence of payment of allowance to parent or usurpation of property by fraud. Loneliness and isolation are two most important worries for an old person. Indian law bypasses this and addresses only material concerns of the elderly. Counter arguments of the new Chinese law to a certain extent does makes sense as

> Kinship is part of human nature; it is ridiculous to make it into a law. It is like requiring couples who have gotten married to have a harmonious sex life (*The New York Times*, 2013).

Yet many would argue that without this kind of law, neglect will continue unchecked. With it at least some parents would benefit. However, any such Act should specify clearly what it means by frequent visits, clauses under which it can permit the elderly to lodge a complaint against the child and associated sanctions for not frequently meeting the parents. A comparative framework between India and China does not aim to gloss over the peculiar situation in China accruing to the One-Child Policy (enacted in 1980). This leaves the current generation with less people to be count upon in old age.[14] A bigger concern arises in the context of

[13]This is how the MW Act defines 'maintenance'.

[14]http://www.usatoday.com/story/news/world/2013/07/01/china-children/2480593/.

'shidu' families, *shidu* in Chinese refers to 'loss of only child'. This makes it morally necessary in the case of China to legalize care or what in Indian language would mean *seva.* When the problem of lonely parents is becoming grave in India, one may wonder if a similar law (as in Chinese context) can provide some respite. Community assistance through locality clubs can bring in greater engagement with such lonely elders. Also, greater weightage could be given to community service for elderly for the purpose of getting any government benefit and employment. The local government has a big role to look at this cultural aspect.

Law's Construction of the 'Docile' Body of the Senior Citizen

The location of juridical care, which is care that is administered, planned, monitored and justified by the State through its policies is essentially located in the body of the elderly. Let us take the example in *Lotika Sarkar v. Preeti Dhoundial and Ors* (2009). In this case, during the hearing, it was alleged that LS had willingly signed the gift deed of transfer of property to the D family. LS agreed that she had read and understood the contents of it after which she signed. However, she came to know that the gift deed she had signed in reality meant a complete transfer of her property rights to the D family, only when there was a media furor about it, after the D family disallowed LS's domestic maidservant to stay back in the house. The court ruled that she signed it without having understood the implication of it in terms of complete property transfer. The gift deed was nullified on accounts that the plaintiff LS was in her 'ripe age', with a 'fragile body' and 'failing memory'. A gift deed is an unconditional deed, and it is only in the MW Act that a gift deed can be revoked. The language of care that was operationalized by the State, through cancelling the deed and going against the D family, was localized in LS's bodily conditions of the 'fragile body' and 'failing memory' discourse.

The extent of resistances to injustices experienced by the right-bearing senior citizen at the hand of children also depends on the way law itself can work reflexively and improve upon itself. For example, some legal provisions discriminate against parents, for instance parents not being considered for entitlement to family pension in case of death of child in government jobs as in the case of *Geeta Devi v. Financial Commissioner & Principal Secretary to Govt. Haryana and Others* (2013).[15] In this case, the Haryana Aided Schools (Special Pension & Contributory Provident Fund) Rules, 2001 under Rule 14 'dependent family members' did not include parents (here GD, the petitioner) to be considered for pension. However, GD's advocate took recourse to precedent cases under the Punjab Civil Services Rules in which parents were included in the definition of family under Rule 6.16B for payment towards death-cum-retirement gratuity. Parents were also included in family under Rule 8.34 for entitlement to grant of

---

[15]MANU/PH/0628/2013.

'wound and other extraordinary pension'. However, parents were excluded from definition of family under Rule 6.17 for grant of family pension. Thus, the Division Bench clarified that there was no logic of such exclusion and held it to be 'arbitrary'. 'Next to God, thy parents, says the poet' was quoted from the same judgement. The court ordered for a redefinition of family under the same Rule. GD's case thus referred to this prior judgement to ensure that the government-aided school in which her deceased son worked would be coerced towards payment of the family pension to the dependent mother. The above case depicts a contradiction where, on the one hand, the State Rules exclude the elderly parents from definition of family; and on the other side, the higher court orders for them to be included within the definition of family, to be eligible for family pension. Thus, the elderly is made a 'dependent', and often only by stereotyping a person as a dependent, benefits can be availed. Thus, vulnerabilities and stereotypes are created out of law and are sustained through legal discourse. They seldom get erased from the system.

The Practice of Care at a Legal Advocate's Chamber

Old age problems that the care providers—State and non-State actors like non-governmental organizations (NGOs)—conceptualize primarily emphasize material aspects of care like property disputes. This became evident when I interviewed the legal advocate who sits every Monday at a renowned NGO's Age-care Centre in Delhi. The well-groomed advocate sat at a corner in the office space of the NGO. To demarcate the privacy, a curtain had been put up so that people in the adjacent space designated for physiotherapy would maintain a boundary from the troubled clients. The advocate explained that most of the clients were elderly persons who faced property issues. They came there either alone or with some concerned relatives. They were mostly nervous and 'shiver' as she described. So she gives them a 'patient hearing and with a calm and composed attitude' and asks them details about the problem they are facing. The primary motive of the lawyer is to instil courage in the elderly person to be able to take up the issue with confidence.

The advocate informed me that complicated cases are addressed through a Mediation Centre. Initially, reconciliation at the level of NGO is sought through a joint hearing in order to reach an agreement 'as per the desire of the parties' to quote the advocate. This takes the form of family counselling. She says that most often the opponent party who has allegedly abused an elderly knows that it is wrong. Thus, several cases get resolved through such counselling. Even after counselling if the matter is not resolved, then it is taken to court to be heard by the Maintenance Tribunal.[16] In the middle of the conversation, the advocate asked if I

[16]The clients can directly approach the Maintenance Tribunal, Women's Commission with an application against the relative. In this chapter, I discuss an alternative way of grievance redressal of the elderly through the legal aid provided by an NGO.

could fight a case in the Tribunal for the elderly who was presenting his case to the advocate. She told me that I could 'bring justice to an elderly'. Initially, I did not wish to upset the situation by telling the man that I had no expertise or experience with courts. However, when the advocate insisted I had to politely inform that due to professional constraints, I would not be able to give my consent immediately. At this, she lost her cool and asked me to leave. She seemed disinterested in talking to me any further. For the advocate, my refusal indicated an unwillingness to provide help or moral support to the elderly that should be provided to them when cases drag to the court. However, the question of legal representation came up through this episode with the advocate. According to the MW Act, the aggrieved elderly can claim maintenance, but in subsequent proceedings of the case, lawyers are not allowed to represent them. One could note this in chapter II of the Act under sections 17 and 18 that no parties are entitled to representation by a legal practitioner. The State government can appoint a Maintenance Officer for their representation in courts only if the parent wishes so. This has a good intention to prevent profit-seeking legal agents to take advantage of often emotionally disturbed elderly. However, in many cases, as this one, the elderly were left by themselves and often rely on the help of random people who may help them out of sympathy. It is also questionable as to what extent the Maintenance Officer would be available to effectively attend to every case. There could be several ongoing cases at the same time. The personal encounter highlighted the lack of a systematic framework in place and existence of a rather loose system of addressing concerns of elderly persons in the country. It was also not clear why the legal advocate who is willing to represent the elderly cannot do so if she wishes to take it up on pro bono basis. Also, there is no legal provision which addresses emotional torture and abuse specifically to the elderly. If such cases arise, the local police is supposed to 'solve' the matter.[17] On the whole, there seemed to be a dearth of connection points between the aggrieved elderly and legal system, mostly due to lack of awareness.

Conclusion

One significant point that comes through an analysis of the legal and policy discourse in India is that such discourse cannot be dissociated from the discourses at the international domain. The establishment of the discipline of gerontology and the phases of discursive shifts are important if one is to locate the certain laws and policies enacted in India. Also the paper finds that the two kinds of care, viz. the juridical care and the medicalized model of care, are not completely disjointed. Many of the State's provisions cater to the very technically treatable bodies of the elderly.

The power that the legal judgements, the various policies and schemes have in constructing the identity of the elderly persons can be seen in the way it marks the

---

[17]I was informed about it by the NGO worker where I encountered the advocate. By 'solving', she hinted at compromise and reconciliation between the abuser and elderly.

body of the elderly citizen as docile. Docility is inscribed as a vulnerable condition. The various judgements of the courts show that it views the elderly as vulnerable. The language in which the 'active'ness is conceived is with respect to the ability of the elderly to perform creatively and productively in their lives (NPOP 1999). There is an implication through the policies, laws and judgements that the ability to perform by staying 'active' is desirable condition but no attempt is made to consider that 'active' can subjectively vary across people and cultures. The governmental category of care that is constructed through the legal discourse is sufficiently medicalized, with an identification of the difference between the normality of the body through staying active and the otherwise 'ageing' body. There is no challenge brought about by the policies to the concepts that are universalized through international gerontology, for instance the idea of 'active ageing' is itself not clearly dealt with in the Indian context and what it could imply in different regions, especially rural and urban. Thus, in these ways as discussed above, the role as a senior citizen, empowered through rights, yet docile, is being performed and institutionalized through legally scripted transactions.

Much of the hierarchy between the care provider and the caregiver is institutionalized through a discourse about the healthy body which is 'normal'. It is conveniently forgotten by policymakers that 'normality is indeed normative' and normality is an evaluation rather than an expression of mathematics (Canguilhem 1991). Thus, there could be no objective criteria of normal healthy aged in the first place. I argue on the lines of Sevenhuijsen (1998) that 'vulnerability', 'fragility' and 'dependency' are part of human existence. One must, according to Sevenhuijsen (1998), refrain from making binary categories, for instance if one is to receive care, it means that the person is not autonomous. According to her, autonomy and care can be two sides of the same coin. Likewise, one's autonomy can be a function of the way the person maintains caring relationships with others around her/him.

The State engenders its own discourse on care which translates as kind of material care, which includes the 'maintenance' for the sustaining the body of the elderly. Care, in State's language seldom, addresses the aspect of psychological well-being. Law has the potential to be an important agent of change in society with respect to care for elderly in accordance with the social changes in India. The transition of care function from the household members to the others—agencies, old age homes—should be made through provisions within the law, so as to bring increased identification and dignity to care work. The lack of a regulatory framework for care workers in the country is a significant lacuna. Law could be crafted in a way so as to accord dignity to care work, prevent exploitation in care-related works and ensure rights of the workers. The juridical dimension of care centres on the strategic deployment of persons as right-bearing citizens. The identity of the elderly is 'performed' through the State's reiterative nomenclatures and language of care which locates empowerment through reinforcing the notion of the passive ageing body. Thus, care in juridical terms becomes a service to be given to the 'needy', situated in a disadvantaged situation than many others in terms of various resources. Care in the juridical discourse obliterates the possibility of viewing the caregiving and receiving as symbiotic and dialogic.

Acknowledgements I am immensely grateful to Dr. Pratiksha Baxi, my PhD advisor at Centre for Law and Governance, J.N.U., without whose inspiration and insights my analyses would have remained incomplete.

References

Bose, A., & Shankardass, M. K. (2004). *Growing old in India*. Delhi: B.R.Publishing Corporation.

Boucher, G. (2006). The politics of performativity: A critique of Judith Butler. *Parrhesia, 1*, 112–141.

Brijnath, B. (2014). *Unforgotten: Love and the culture of dementia care in India*. New Delhi: Orient Blackswan Pvt. Ltd.

Butler, J. (2007). *Gender trouble*. Great Britain: Routledge.

Canguilhem, G. (1991). *The normal and the pathological*. New York: Zone Books.

Cohen, L. (1998). *No aging in India: Modernity, senility and the family*. UK: Oxford University Press.

Donzelot, J. (1997). *The policing of families*. USA: John Hopkins University.

Feinberg, J. (1971). Legal paternalism. *Canadian Journal of Philosophy, 1*(1), 105–124.

Foucault, M. (2003). *The birth of the clinic*. London: Routledge.

Fraser, N. (1989). *Unruly practices: Power, discourse and gender in contemporary social theory*. Minneapolis: University of Minnesota Press.

Government of India (GOI), Section 125 of Criminal Procedure Code of 1973.

Government of India (GOI). (2007). *Indira Gandhi National Old Age Pension Scheme*. National Social Assistance Programme (NSAP), Implemented by Ministry of rural development.

Government of India (GOI). (2008). *Integrated programme for Older Persons (IPOP)*, Revised Scheme. Ministry of Social Justice and Empowerment.

Green, B. S. R. (1993). *Gerontology and the construction of old age*. New York: Walter de Gruyter Inc.

Katz, S. (1996). Disciplining old age: The formation of gerontological knowledge. USA: The University Press of Virginia.

Kleinman, A. (1988). *The illness narratives: Suffering, healing and the human condition*. USA: Basic Books.

Liebig, P., & Rajan, S. I. (2005). *An aging India: Perspectives, prospects and policies*. Jaipur: Rawat Publications.

National Policy on Older Persons. (1999). *Ministry of social justice and empowerment*. New Delhi: Govt. of India.

National Programmes for the Health Care of the Elderly (NPHCE). (2010–11). *Ministry of health and family welfare*.

National Policy for Senior Citizens. (2011). *Ministry of social justice and empowerment, government of India*.

Paltasingh, T., & Tyagi, R. (2012). Social security & policy on senior citizens: evidences from Asian countries. *Ageing & Society, Indian Journal of Gerontology, 22*(1&2), 1–20.

Political Declaration and Madrid International Plan of Action on Ageing 2002. New York: United Nations.

Sevenhuijsen, S. (1998). *Citizenship and the ethics of care: Feminist considerations on justice, morality and politics*. London and NY: Routledge.

Soneja, S. (2011). *Elder abuse in India*. HelpAge Country Report for WHO.

The Maintenance and Welfare of Parents and Senior Citizens Act. (2007). *Ministry of law and justice*.

Wong, E. (2013). A Chinese virtue is now the law. *The New York times*. http://www.nytimes.com/2013/07/03/world/asia/filial-piety-once-a-virtue-in-china-isnow-the-law.html?_r=0. Retrieved June 22, 2016.

Zola, I. K. (1983). *Socio-medical inquiries*. Philadelphia: Temple University Press.

Cases

Lotika Sarkar v. Preeti Dhoundial and Ors. (2009). MANU/DE/3216/2009.
Anbalagan v. The Sub Divisional Executive Magistrate and Revenue Divisional Officer. (2012) W.P.No.12145 of 2012.
Wp(C). No. 17245 Of 2012 (E) vs By Adv. Sri. Varghese P.Chacko (http://indiankanoon.org/doc/170672315/).
Geeta Devi v. Financial Commissioner & Principal Secretary to Govt. Haryana and Others. (2013) MANU/PH/0628/2013.

Websites

http://timesofindia.indiatimes.com/india/Abuse-on-elders-increase-by-50-this-year-Helpage/articleshow/36503113.cms. Retrieved August 12, 2014.
http://www.usatoday.com/story/news/world/2013/07/01/china-children/2480593/. Retrieved May 2, 2015.
http://timesofindia.indiatimes.com/business/india-business/Unemployment-levels-rising-in-India-experts-say/articleshow/29403619.cms. Retrieved August 12, 2015.
http://blacks.worldfreemansociety.org/1/J/j-0663.jpg. Retrieved November 5, 2015.
http://www.businesstoday.in/moneytoday/real-estate/old-age,-new-home/story/6258.html. Retrieved February 2, 2016.

Chapter 4
Ageing Population: Need for Formal Care Workers

Bharti Birla

Abstract The world population is ageing, and the number of elderly needing care and support will grow further in coming years. The challenges are twofold. Firstly, most of the elderly fall outside the social protection net (e.g. they do not have pensions, healthcare insurance, savings), which means that it is difficult to pay for care services. Secondly, there is a general lack of state provision of care services for the elderly. In most cases, the family provides such care and support. Most often, the women of the household discharge this responsibility. When families can afford, they can take services of the paid care workers, freeing the women in the household to take up paid employment. The feminized nature of care work, undervaluation of women's work in general and lack of legislations that recognize and regulate paid domestic work and care work, often pushes care workers in precarious employment relationships with poor working conditions and low wages. The demand for paid care elderly care, especially long-term care and support, is increasing, but there is an acute shortage of professionally skilled domestic workers who can provide elderly care services.

Keywords Elderly · Ageing population · Care workers · Domestic workers · Social protection · Skilled workers · Feminization of care services · Paid employment

---

The views expressed are those of the author and do not necessarily reflect those of the International Labour Organization, or of the ILO DWT for South Asia and Country Office for India. In case of suggestions or corrections, please write to birla@ilo.org.

B. Birla (✉)
International Labour Organization, DWT for South Asia and Country Office for India, 3rd Floor, Core 4B, Habitat Centre, Lodhi Road, New Delhi 110003, India
e-mail: birla@ilo.org; bharti_birla@hotmail.com

S. Irudaya Rajan and G. Balagopal (eds.), *Elderly Care in India*,
DOI 10.1007/978-981-10-3439-8_4

Introduction

Care work is broadly defined by the ILO as the work of looking after the physical, psychological, emotional and developmental needs of one or more other person(s). Care recipients are generally identified as infants, school-age children, people who are ill, persons with a disability and elderly people. Care providers typically include public and private health services, state-regulated or public-sector social workers, public or private care-provider agencies, enterprises of employment, voluntary and community organizations, faith-based organizations or networks and relatives and friends. Different settings and modalities of care work apply to each of these categories (ILO 2007).

Gendered Notions of Care Work

Though both men and women are involved in care work, it has been seen that unpaid household and care work falls disproportionately on women. Having to assume responsibility for care work within families, women tend to work longer each day. This impacts women's opportunity to join full-time paid work (employed or self-employed), augmenting the existing labour market inequalities in the context of women and work. This also contributes to lower rates of female labour force participation and occupational and sectorial segregation, often pushing women towards lower quality of jobs, widening the income gap, slowing career progression and reducing social protection cover (ILO 2016). Not only at the workplace, unpaid care work also leads to discrimination within the household, as women's work is undervalued and unrecognized. Women have to juggle between paid work, family responsibilities and personal life. Lack of care services (private or public) providing good quality, affordable and accessible childcare and other social care services for family members combined with lack of family-friendly, flexible working arrangements pushes women, who are often paid lower than their male counterparts, to leave the labour market either temporarily or permanently (ILO 2016). Even after a short break, it may be difficult for women to continue working without suffering a penalty (Aisenbrey, Evertsson, and Grunow 2009). For instance, in a 2014 poll conducted in the USA, 61% of women said that they were not working because of family responsibilities. Three quarters of homemakers would consider working if they could work flexible hours or work from home (Miller and Alderman 2014).

In developed countries, women spend an average of 4 h and 20 min on unpaid care work, while men spend 2 h and 16 min per day. In developing countries, women spend 4 h and 30 min per day on unpaid care work compared to 1 h 20 min for men. The reasons for these gender gaps are related to the lack of services and infrastructure to reduce unpaid household and care work, along with gender-based and social norms under which women are considered to be the primary care providers (ILO 2016). And not just women, girl children (below 14 years) and adolescent girls (14+–18 years) also need to drop out of school to take care of their younger siblings when their parents go for work. Gender discrimination in context of care workers is intertwined with the gender norms, societal expectations and patriarchal family structures. The women in the family are expected to provide care services to the elderly, while the needs of women themselves, who provide this 'family care', are ignored. Such women are deprived of opportunities to join paid work and secure their own future in terms of income, associated social protection which they have a potential to receive as a paid worker. In some countries, like India, the law for the protection of the elderly mandates the children to take care of the aged parents (GoI 2011), including their financial and healthcare needs, with no economic contribution from the state. This kind of forced 'family care' by female members of the family aggravates existing gender gaps.

Ageing Population and Need for Care Work

In 1950, the world population aged 60 years and above was 205 million (8.2% of the population) which increased to 606 million (10% of the population) in 2000. By 2050, the proportion of older persons 60 years and above is projected to rise to 21.1%, which will be two billion in number. Asia has the largest number of world's elderly (53%), followed by Europe (25%). In 2050, 82% of the world's elderly will be in developing regions of Asia, Africa, Latin America and the Caribbean, while only 16% of them will reside in the developed regions of Europe and North America. Population ageing is therefore rapidly emerging as the problem of developing countries (Raju 2011).

With only 5.6% of the global population living in countries that provide universal long-term care coverage through social protection, and more than 48% of the world's population not protected by national legislation to this effect, and another 46.3% are, to a large extent, excluded from long-term care coverage due to narrow regulations that limit benefits only to the poorest (Xenia Scheil-Adlung ILO 2015), the long-term care for the elderly is becoming a major challenge (Map. 4.1).

There is a huge unfulfilled demand for care workers for the elderly, especially for long-term care. Filling these gaps would create employment—particularly for women and in rural areas where gaps are most severe—and provide access to urgently needed services. The absence of formal care workers results in the exclusion of large parts of the older population from quality services. Most severe shortages for long-term care are found in Asia and the Pacific, which needs 8.2 million care workers as 65% of the population (representing 300 million people) aged 65+ remain without formal long-term care services (Xenia Scheil-Adlung ILO 2015). And this demand for long-term care workers is expected to significantly increase in the future due to demographic ageing. The ILO study (Xenia Scheil-Adlung ILO 2015) covered 80% of the world population aged 65 and over found extreme deficits in social protection for older persons in need of long-term care due to a lack of 13.6 million long-term care workers worldwide. And what is worrisome is that these shortfalls exist despite the fact that the bulk of care—up to 80% of long-term care work—is provided by unpaid female family members of older persons.

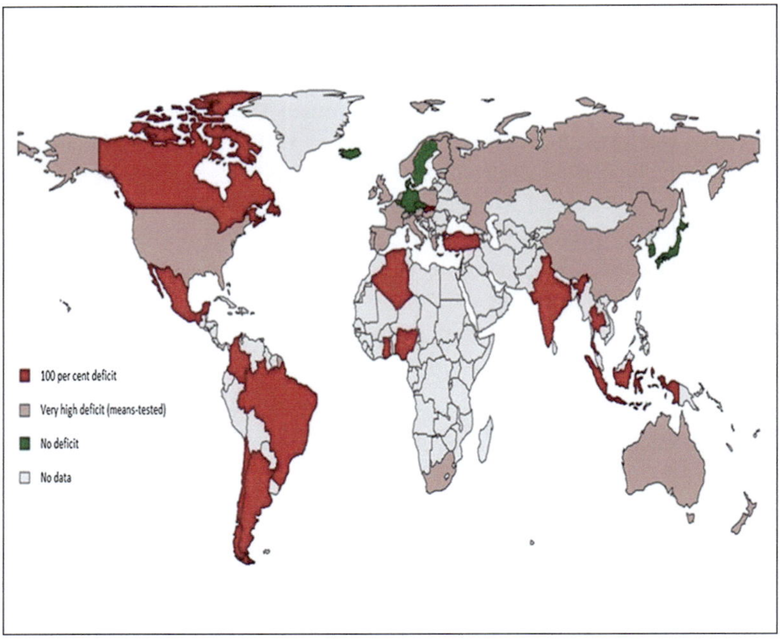

Map. 4.1 Deficits in legal LTC coverage towards universal coverage based on national legislation, 2015 (total population, percentages). *Source* ILO estimates 2015, World Bank, 2015 (population data in 2013)

The ILO's work on changing demographics can offer guidance as to the size and scope of the problem of the elderly and need for skilled elderly care workers. By 2050, the world's population will surpass 9 billion, and the number of people aged over 60 years and over will have tripled see (ILO 102nd ILC Session, IV Report 2013). Most of these people will require a wide range of care services, not only health and medical care, but also everyday personal and household care. Further, preventive care is hardly being provided to the elderly, as the potential of capacity improvements of older persons is often neglected. There is also little appreciation of the fact that positive developments can be achieved by providing adequate quality long-term care services to elderly.

Paid Care Workers: Situating Domestic Work in Care Work

Domestic work is defined as work for and in a household or households. The employment of women domestic workers belonging to socioeconomically disadvantaged groups for care work that caters to the needs of higher socioeconomic status groups, while underplaying their own care requirements, has a long history (Razavi 2007). A majority of domestic workers provide personal and household care; they may cook, clean and take care of children, the elderly and the disabled in private homes. Due to lack of state provisions of care services, families resort to paid care work, largely by recruiting domestic workers. Domestic workers work for private households, often without clear terms of employment, mostly as unregistered workers. The domestic work itself remains unregulated and mostly excluded from the scope of labour legislations in a large number of countries. According to ILO estimates (ILO 2013), there are at least 53 million domestic workers worldwide, not including child domestic workers, and this number is increasing steadily in developed and developing countries. The gendered notions of care work are visible in domestic work too. Globally, one in every 13 female wage earners is employed in domestic work. This is a highly feminized sector, though the substantial number of men also works in the sector. Most men are employed in better paying domestic chores such as gardeners, drivers or cooks. About 83% of all domestic workers are women.

A domestic worker may work as a live-in worker (residing in the household of the employer) or a full-time (works full day for the same employer, usually live-out worker) or may work on a part-time basis, where she/he may be employed by a single household or by multiple employers. A domestic worker may be working in a country of which she/he is not a national or come from source areas within the country to work in urban centres as inter- or intra-state migrants. In these cases, usually, they are referred to as migrant domestic workers. At present, domestic workers often face very low wages, excessively long hours, have no guaranteed weekly day of rest and at times are vulnerable to physical, mental and sexual abuse or restrictions on freedom of movement. Exploitation of domestic workers can partly be attributed to gaps in national labour and employment legislation, and often reflects discrimination along the lines of sex, race and caste. It is estimated that US$ 8 billion are saved by private households by not paying or underpaying domestic workers held in forced labour (ILO 2014). Based on information in the ILO's 2012 Database on Global Estimate of Forced Labour, it can be estimated that forced domestic workers are paid on average about 40% of the wage they should receive.

In ageing societies, family safety nets are breaking and nuclear families are growing. The public health deficits are prominent, and even where they exist, they are shrinking. The provision of the care services by the state is limited and more concentrated in the developed world. All this is leading to a greater demand for care workers in private households. This demand, to a large extent, is being fulfilled by predominantly female domestic workers. It is also because of them that other women have been able to free themselves from these family care responsibilities to take up paid employment (ILO, March 2015).

Decent Work for Domestic Workers

There are over 53 million domestic workers across the world, and 83% of them are women. The caregiving role of domestic workers enables other women to enter the labour force, allowing them to reconcile work and family responsibilities (ILO, March 2015). The role of domestic workers is increasingly being recognized across the globe, especially after the adoption of the ILO Domestic Workers Convention (C189) on promoting decent work for domestic workers. However, domestic workers still remain at the lowest end of the care economy. The exclusion from labour laws or weak laws for regulating the wages and working conditions of domestic workers, as well as low dignity assigned to the profession, is an indicator of the unequal treatment meted to this feminized labour force. A majority of these workers are migrants who are often undocumented and performing their work under unclear terms of employment. Most temporary migrant workers do not enjoy either the right to become permanent residents in their host country or the right to circulate freely in the labour market. Societies and families place low value to unpaid care work that women undertake. Consequently, assuring decent work for domestic workers establishes the principle that domestic workers, like any other workers, are entitled to a minimum set of protections under labour law (ILO Global Statistics 2013).

Professionally Skilled Elderly Care Workers

Elderly care work is a specialized type of domestic work that needs worker to have the right skills as well as aptitude to undertake the work. The needs of the elderly differ from younger people, and working for elderly means that domestic workers should have compassion, patience and expertise to undertake elderly care work. The common problems associated with the ageing process such as the change in behaviour pattern of the elderly, the associated old age illnesses and effects of medicines can all make the work of an elderly care worker more challenging. Men and women age differently, and hence, their requirements are different. These gender differences in the ageing process and the different needs of aged men and women need to be understood. For meeting this continuum of care for the elderly, pain and palliative care, care with toiletry for the elderly, taking care of persons with Dementia/Alzheimer, handling terminally ill elderly and providing long-term care for the elderly require professional vocational skills training.

In many countries, there are well-developed skills training programs which enable men and women to get trained as a skilled elderly care worker. These training programmes use modular, skill-based courses that recognize prior learning and enable trainees, mostly women, to work professionally as an elderly care worker, childcare worker, household cook, etc. This type of skilled training provides domestic workers a definite career path, which can take them gradually to the hospitality or para-health industry. In India, the government of India has already approved a Domestic Workers Sector Skills Council, which is responsible for establishing a national institutional network for skill development of domestic workers. Different vocational training providers affiliate with the council to deliver skills training. The Council has become operational in 2016.

Skills is one of the key tools to promote decent work as it facilitates better wages and working conditions, and enables an improved environment for both the workers and the employers. When planning for skilling and placement of domestic workers, it is important that the skill providers and their trainers also understand the nuances of domestic work—work like any other, work like no other. Domestic work is characterized by highly individualized employment relationship, with levels of control by the employer, high levels of job insecurity, poor working conditions, unequal power relations, unregulated and peculiar nature of work, poor career growth, lack of privacy of the workers and treatment as unskilled/low skilled work.

In this context, it is important that skills for domestic workers are not just seen as technical skills, but a package of soft and life skills embedded in the rights-based, technical curriculum to enable domestic workers to negotiate the complex employment relationship and achieve better employment outcomes in unregulated and peculiar work conditions.

Conclusion

The care economy offers an opportunity for paid domestic work and has enormous potential for employment generation in the coming years. It will be necessary to strengthen it to continue to improve gender equality for all women, including domestic workers. It is evident that home care workers and elderly care workers will be among the fastest growing occupations. Therefore, it is important to find solutions that address the care economy, such as provision of state-based public care services which can be accessed by a number of people in a collective manner. The access to social protection for all, including care workers, is most crucial as the population grows. Investing in care services would also enable equal opportunities for women in the world of work. Though men in developed nations have shown an increase in time spending on care work, sharing some burden of the women, it is still not equal (ILO 2016). Enabling the care burden to be shared between the state, employers, paid elderly care workers, unpaid men and women as family caregivers is important.

References

Aisenbrey, S., Evertsson, M., & Grunow, D. (2009). Is there a career penalty for mothers' time out?: A comparison of Germany, Sweden and the United States. *Social Forces, 88*(2), 573–605.

Government of India: Situation analysis of elderly in India. (2011). Central Statistics Office, Ministry of Statistics & Programme Implementation, Government of India.

ILO 102nd ILC Session, IV Report. (2013). *Employment and social protection in the new demographic context.* International Labour Conference, 102nd Session.

ILO. (2007). *ABC of women workers' rights and gender equality.* Geneva: International Labour Organization.

ILO. (2013). *Domestic workers across the world: Global and regional statistics and the extent of the legal problem.* Geneva: International Labour Office.

ILO. (2014). *Profits and poverty: The economics of forced labour.* International Labour Office Geneva.

ILO Global Statistics. (2013). *Domestic workers across the world: Global and regional statistics and the extent of legal protection.*

ILO. (2015). *Women and the future of work—Taking care of the caregivers.* Susan Maybud, ILO Gender, Equality and Diversity Branch (GED), and INWORK Team.

ILO. (2016). *Women at work: Trends 2016.* Geneva: International Labour Office.

Miller, C., & Alderman, L. (2014). *Why U.S. women are leaving jobs behind.* Retrieved from The New York Times (http://www.nytimes.com): http://www.nytimes.com/2014/12/14/upshot/us-employment-women-not-working.html?_r=0.

Raju, S. S. (2011). *Studies on ageing in India: A review,* BKPAI Working Paper No. 2, United Nations Population Fund. UNFPA.

Razavi, S. (2007). *The political and social economy of care in a development context: Conceptual issues, research questions.* Gender and Development Programme Paper Number 3, United Nations Research Institute for Social Development.

Xenia Scheil-Adlung. (ILO, 2015). Global estimates of deficits in long-term care protection for older persons/Xenia Scheil-Adlung; International Labour Office. Extension of Social Security series; No. 50.

Chapter 5
Old Age Homes in India: Sharing the Burden of Elderly Care with the Family

Anupama Datta

Abstract Population ageing is silently becoming a major challenge for our country. According to estimates, by the year 2050, 20% of our population will be over the age of 60 years. Average life expectancy at the age of 60 years is also expected to increase to 30 years from the current rate of 17 years. This coupled with the fact that the adding years to life may not always mean adding life to years. In other words, advancing age brings with it the threat of frailty, dependence and handicap. The demographic estimates also say that the number of younger people is likely to decrease due to lower birth rate. So, care in old age that was so heavily dependent on family will be in peril. The decreasing family size, immigration and changing ethos have started posing challenges for care in many parts of the country. The elderly can no longer blindly depend on family to take care of them in the twilight years. Old age homes are filling the gap to a certain extent. But as of now, it is not a popular and affordable choice for most. Traditionally, old age homes were meant for the poor and destitute and hence mostly managed by charitable organisations, but in recent years, paid facilities have also emerged to cater to the needs of middle and upper middle class older persons, who can pay for care in old age. However, due to its unstructured growth, there is no regulatory regime to deal with it. This is a cause for concern in more than one way.

Keywords Old age homes · Care · Family · Charitable organisations · Paid facilities

---

A. Datta (✉)
Policy Research and Development, HelpAge India, C-14 Qutab Institutional Area, New Delhi 110016, India
e-mail: anupama@helpageindia.org

S. Irudaya Rajan and G. Balagopal (eds.), *Elderly Care in India*,
DOI 10.1007/978-981-10-3439-8_5

Background

It is a known fact that the number of older persons in India is increasing and the pace is rapid. It is not enough to acknowledge the fact, but important to go deeper into it and understand the nuances in terms of gender, age, area, health, independence, etc., as that will enable an understanding of the needs of each sub-segment of the older population to deal with it effectively. A quick glance at the population projection figures at the outset will be useful. According to the estimates of Government of India, in the first 25 years of twenty-first century, the population of older persons will increase from 6.9% (7.06 crore) of the total population in 2001 to 12.4% (17.31 crore) in the year 2026. In the same period, the life expectancy of males will increase from 63.8 to 69.8 years and that of females from 66.1 to 72.3 years.

If we look at age disaggregation, we find that in the year 2001 the young-old constituted 61.76% of the elderly population and this is expected to decrease to 58.36% in the year 2026. But the population in the other two segments old-old and oldest-old are likely to increase. The old-old will increase from 27.74 to 30.15% and oldest-old from 10.49 to 11.47% in the same period. In other words, the older population is getting older. The gender disaggregated data show trends towards feminisation of ageing population.

Experts have proved and commoners know that ageing brings with it inevitable declines in physical and mental capabilities. It may vary from person to person and segment to segment, but the bottom line is that as people age, they become frail and dependent. This ageing population whatever its composition requires care and shelter as it is one of the basic needs for a life with dignity. Traditionally, these needs were being fulfilled satisfactorily by the family. However, family is still the main source of support for the older persons in India; but this source is gradually diminishing and the apprehension is that in few decades from now, it will not be able to shoulder the responsibility. The apprehensions are based on the current trends of family size, orientation, migration, longevity, increasing morbidity with age, cost of living, poverty and impoverishment and many more.

The National Policy on Older Persons 1999 (NPOP), which is the national instrument on ageing issues, covers these aspects besides whole host of other equally pressing issues like financial security, health. It is useful to highlight the provision of the NPOP on care and shelter to understand our national perspective on the issues. The NPOP document in Point 11 gives one of the implications of population ageing thus:

> Industrialisation, urbanization, education and exposure to life styles in developed countries are bringing changes in values and life styles. Much higher costs of bringing up and educating children and pressures for gratification of their desires affects transfer of share of income for the care of parents. ***Due to shortage of space in dwellings in urban areas and high rents, migrants prefer to leave their parents in their native place. Changing roles and expectations of women, their concepts of privacy and space, desire not to be encumbered by caring responsibilities of old people for long periods, career ambitions, and employment outside the home implies considerably reduced time for care giving.***

> Also, adoption of small family norms by a growing number of people implies ***availability of fewer care givers specially since in a growing number of families, daughters, too, are fully occupied, pursuing their educational or work career. The position of single persons, particularly females, is more vulnerable in old age as few persons are willing to take care of non-lineal relatives. So also is the situation of widows an overwhelming majority of whom have no independent source of income, do not own assets and are totally dependent.***

Further in Point 19 of NPOP Statement, we find:

> The Policy values an age integrated society. It will endeavour to strengthen integration between generations, facilitate two way flows and interactions, and strengthen bonds between the young and the old. ***It believes in the development of a social support system, informal as well as formal, so that the capacity of families to take care of older persons is strengthened and they can continue to live in their family***.

One of the Principal Areas of Intervention and Action Strategies mentioned in NPOP is Shelter and Point 48 of the Policy document states that:

> ***Shelter is a basic human need***… Housing schemes for urban and rural lower income segments will earmark 10% of the houses/house sites for allotment to older persons. This will include Indira Awas Yojana and other schemes of government. Earning persons will be motivated to invest in their housing in their earning days so that they have no problems of shelter when they grow old. This will require speedy urban land development for housing, time bound provision of civic services and communication links, availability of loans at reasonable rates, easy repayment instalments, time bound construction schedules and tax reliefs. Development of housing has to be a joint endeavour of public and private sectors and require participation of Housing Development Boards, civic authorities, housing finance institutions and private developers and builders. ***Older persons will be given easy access to loans for purchase of housing and for major repairs, with easy repayment schedules.***
>
> Layouts of housing colonies will have to respond to the life styles of the elderly. ***It will have to be ensured that there are no physical barriers to mobility, and accessibility to shopping complexes, community centres, parks and other services is safe and easy….*** It will therefore, be necessary to earmark sites for such centres in all housing colonies. ***Segregation of older persons in housing colonies has to be avoided as it prevents interaction with the rest of the community….***Preferences will be given to older persons in the allotment of flats on the ground floor (Point 49, NPOP 1999).
>
> ***Group housing of older persons comprising flatlets with common service facilities for meals, laundry, common room and rest rooms will be encouraged.*** These would have easy access to community services, medicare, parks, recreation and cultural centres (Point 50, NPOP 1999).
>
> Older persons and their families will be provided access to information on prevention of accidents and on measures which enhance safety, taking cognizance of reduced physical capacity and infirmities (Point 51, NPOP 1999).
>
> ***Civic authorities and bodies providing public utilities will be required to give top priority to attending complaints of older persons….*** (Point 54, NPOP 1999).

Another area listed under Principal Areas of Intervention and Action Strategies in NPOP that is relevant and important is **Welfare**. The provisions are as follows:

> The main thrust of welfare will be to ***identify the more vulnerable among the older persons*** such as the poor, the disabled, the infirm, the chronically sick and those without family support, ***and provide welfare services to them on a priority basis. The policy will be to consider institutional care as the last resort when personal circumstances are such that stay in old age homes becomes absolutely necessary*** (Point 59, NPOP 1999).

> ***Non-institutional services by voluntary organizations will be promoted and assisted to strengthen the coping capacity of older persons and their families. This has become necessary since families, as they become smaller and women work outside the home, have to cope with scarcity of full time care givers. Support services will provide some relief through sharing of the family's caring responsibilities (Point 60, NPOP 1999).***

> ***Assistance will be provided to voluntary organizations by way of grants-in-aid for construction and maintenance of old age homes. Those for the poor will be heavily subsidized.....Non-governmental organizations will be encouraged to seek professional expertise in the designing of old age homes***, keeping in view needs of group living at this ***stage of the life cycle*** and the **class of clients** they serve. ***Minimum standards of services in such homes will be developed and facilities provided for training and orientation of persons employed in these homes*** (Point 61, NPOP 1999).

> ***Voluntary organizations will be encouraged and assisted to organize services such as day care, multi-service citizen's centres, reach-out services, supply of disability related aids and appliances,*** assistance to old persons to learn to use them, ***short term stay services and friendly home visits by social workers. For old couples or persons living on their own, helpline, telephone assurance services, help in maintaining contacts with friends, relatives and neighbours and escorting older persons to hospitals, shopping complexes and other places will be promoted for which assistance will be given to voluntary organizations.....*** (Point 62, NPOP 1999).

> The need for plurality of arrangements for welfare services is recognised. ***Government, voluntary organizations and private sector agencies all have a place, the latter catering to those who have the means and desire better standards of care*** (Point 64, NPOP 1999).

It can be implied from the above-mentioned provisions of NPOP that in the government policy, shelter and care are considered important aspects of needs of ageing population and requirement of appropriate intervention at all levels is also clear.

Another instrument that needs mention here is the Maintenance and Welfare of Parents and Senior Citizens Act 2007. The broad analysis of the Act makes the division of responsibility for shelter and care needs of older person between the family and the government. It is clear that the law divides the responsibility of shelter and care for older persons between the family and the government—the economically comfortable family to take care of the older members in case of want and the government to open old age homes to support poor and destitute. On deeper analysis, it will appear as an acknowledgement of need for institutional care in old age. It is not mandatory for the well-to-do adult children or descendants to keep older parents in the family, but to provide for their life with dignity. They provide maintenance allowance. The older persons could use the money to avail paid services for care.

The inability and/or unwillingness of either party (old parent(s) and children) to live with the other as a family with obligations of care, respect and love has increased the need for institutional arrangements. Before we examine in detail the shelter and care needs of the elderly in India and the institutional and non-institutional care structures, it is also useful to pose the broad questions that need to be addressed in this context. The most important questions are as follows: What are the immediate needs of the older persons? What will be their needs in the near future? What kind of facilities, home care and/or institutionalised care, will be required? Who are the main providers: private, charity, public, and what kind of role is envisaged for the future? What are the basics minimum standards of setting up an old age home? What are the desirable standards? Which regulatory body will oversee the proper functioning of the homes? Who will man the old age homes? Who will train the manpower? What skills are required? Who will ensure safety and security of the older persons in these institutions? Where will they report abuse? What are the laws that will govern such crimes and abuse against the elderly?

To find answers to these questions, we may start by looking at the shelter needs of the ageing population and its changing dynamics in the recent years, particularly after globalisation, liberalisation and marketisation of our economy.

Shelter in Old Age

The number of older persons looking for shelter, other than their traditional family home, is progressively increasing. The reason are numerous, but the most compelling are fewer children, migration, lack of resources, irreconcilable differences in ethos, life style, inability of elderly to build their own house, tendency to shirk responsibility of parent(s), childlessness, failed marriage, loneliness, physical dependence, insecurity, abuse, etc. The grievances reported to various Elders' Helplines, cases registered with the Police and several media stories about abuse, abandonment and cheating related to ownership of property are some of the data sources that clearly point to the increasing tendency towards forcing the elderly to be without *shelter* in the last phase of their life. The current responses to this '*shelterless-ness*' of the elderly have been seen in terms of search for old age homes (OAHs)—paid, partially subsidised or free—in the hope of living the last years of their life with dignity and care.

In India, traditionally, there were old age homes for the destitute built and managed by the private charity or religious organizations. The government and voluntary organisations are recent responders to this need. In a way, the two actors recognised their responsibility towards providing facilities to the poor and/or destitute older persons. Integrated Programme for Older Persons (IPOP) and MWPSC Act of 2007 has established the governmental recognition of this predicament faced by the elderly through incorporation of an obligation on the part of the government to support the establishment of more OAHs, at least one in each district in the country. As per the record, in the year 2015–16, Government of India supported only 238 old age homes and 85 day care centres (DCCs) under the Integrated Programme for Older Persons. Further if we look at the statewise break-up, we find that 136 are in the southern states, i.e. Andhra Pradesh (56), Karnataka (33) and Tamil Nadu (47). The other states where the number of old age homes supported under IPOP crosses double figures are Maharashtra (16) and West Bengal (13). The highest number of DCCs is supported in Andhra Pradesh (25) followed by Tamil Nadu (11), Maharashtra (10), Odisha (10) and Haryana (09).[1] The proposed one old age home per district did not see the light of the day due to lack of funds and lack of clarity on the role of part of central and state governments.

As per the data available with HelpAge India, there are 1280 OAHs in the country. Almost 50% of these OAHs are in the southern states. Elders' Helplines in the metropolitan cities of the region have been receiving calls from the distressed elderly about the violation of original promises by OAH managements, private builders and small house-owners who have converted a portion of their residential property into shelter facilities for the elderly. Builders in Mumbai are said to be fleecing the elderly owners of houses in Group Housing Society in the name of re-development. As this is completely unregulated sector, the cases can only be reported and registered under various other existing laws. Given the old age and consequent handicaps, invoking the laws and taking recourse through the judiciary is a very difficult if not impossible redress for the aggrieved elderly.

Some state governments tried to introduce the concept of public–private partnership (PPP) in the management of OAHs, established by the state. They invited corporate, NGOs with technical resources, Senior Citizens' Associations and other legal entities to manage these institutions with a contribution of recurring and non-recurring grant sanctioned by the state government. But, this has not been very successful, as there are many hurdles in the process due to the differences in work culture of government and non-government organisations.

Some state governments appear to be deliberating upon the need to converge other social protection programmes like Old Age Pension, Old Age Health Insurance, etc., to residents in the OAHs. These are some meaningful initiatives to link the shelter provision with other support systems like economic security, health security and even psycho-social security.

The housing industry in the private sector, like the Health Care/Health Insurance Industry and the Hospitality/Recreation Industry (the three can be grouped together and called as the Elder Living and Care Industry), has begun to see the elderly as 'an ever-expanding class of high-spending consumers' and has rapidly emerged as one of the biggest 'providers' of the elderly population through retirement resorts. These facilities are usually expensive and beyond the reach of older persons requiring care and shelter in old age with their ever-decreasing incomes. In some cases, elderly have also been duped of their money with fancy promises.

[1]http://socialjustice.nic.in/writereaddata/UploadFile/Annexure-VIII.pdf. Accessed on 17th July 2016.

Care Giving for Older Persons Within Family and the Community

'The Idea is to Die Young as Late as Possible'. This is a very wise thought by Ashley Montagu. Most of the people in this world would like this to be the reality of their life. However, very few are fortunate enough to experience it. With increasing age, most people face the challenges of low vision, hearing, mobility, while some others face the challenge of disability due to prolonged illness or accident and/or injury. These challenges not only impact the quality of life of the older person, his/her sense of self-worth, independence and dignity but also the caregivers. In India, in most cases, we find that care is given either by the spouse or by the young adult children, particularly female children (daughter or daughter-in-law) or relatives of the older person. There are very few avenues of institutionalised care for the older persons. In addition, there are still some cultural inhibitions in availing these facilitates. In a nutshell, the responsibility of care of older person is normally shouldered by the family members. In this context, it is important to mention that the expectations from family have remained more or less the same, but its structure has undergone tremendous change in the recent past. This fact is coupled with the fact that longevity has also increased at a rapid pace in the recent years and is likely to increase further. But, it still continues to be the predominant mode of living arrangement for the older persons. If we look at the data, we find that in India, 0.5% older persons living alone live in old age homes/institutions and 4.7% live alone but not in any institution, 12% live with spouse only, 44.8% live with spouse and other members, 32% live without spouse but with children and 3.9% live without spouse with other relatives (NSSO 2006).

In the context of changing demographic scene and social values, the situation of the elderly in Indian society is becoming more critical, and of late abuse and neglect of the elderly in the family context is emerging as a significant problem. There are many micro- and meso-studies to prove its existence and its increase.

The traditional norms and values of Indian society that laid stress on showing respect and caring for the elderly are gradually weakening. The symbiotic and reciprocal relationship between older persons and the family is under threat. Many gerontologists opined that in developed countries, population ageing resulted in a substantial shift in emphasis between social programmes causing a significant change in the share of social programmes going to older age groups. But in developing countries, these transfers take place informally and are accompanied by high social and psychological costs by way of intra-familial misunderstandings and strife. It is because these changes that it can no longer be assumed that the elderly would be able to live comfortably at home receiving care from family members willing to spare time and perform services for them. Given this trend, the elderly face a number of problems to adjust to in varying degrees. These problems range from absence of ensured and sufficient income to support themselves and their dependents, to ill health, absence of social security, loss of social role and recognition, to the non-availability of opportunities for creative use of free time.

Various studies have estimated the rate of abuse prevalent in our country. A review of the few of these studies indicates that the most likely victim of elder abuse are females of very advanced age, role less, functionally impaired, lonely and living at home with someone, primarily her adult child, spouse or other relatives. Studies in India have indicated that more women than men complain of maltreatment in terms of both physical and verbal abuse. The prevalent patterns of elder abuse mainly include psychological abuse in terms of verbal assaults, threats and fear of isolation; physical violence; and financial exploitation. The dependence (economic, physical and emotional) of the older person on the family seems to be directly proportional to the level of abuse faced by the older member. Besides the dependent position of the older person as a risk factor, other factors such as perceived powerlessness, social isolation, drug or alcoholic addiction and antisocial behaviour of the abusers could also be related to elder abuse.

To find effective intervention strategies to deal with elder abuse, it is important to understand and acknowledge the need to help families in difficult situations. For this, we need a holistic family assessment which in the Indian situation would be to study the problem, family system, i.e. its life cycle and environment. A clear understanding of all aspects of the problem is essential before using it as the starting point to develop effective response to it. Some of the aspects that need to be understood are the nature, origin, duration, frequency, the older person's perception of the problem and reaction to it; role of other family member in it; and the reporting and redress.

There is no one way to understand and tackle this problem; but several approaches help us do this. The psychoanalytic approach tries to understand the problems by concentrating on the unconscious forces influencing individual behaviour and its impact on family functioning. The Exchange Framework approach also focuses on the motivations of individuals constituting the family. The basic assumption here is that self-interest is the motivation for all the action of any individual and people are rational calculators of rewards and costs. In other words, individuals in the family with old members must evaluate the profits and cost involved in taking care of the older members. The relationship should appear to more rewarding and less costly for each other. The framework that analyse family as a system-interdependent relationships, it is important to identify the relationships or connections between family members and recognise the implications of these connections since change in any one member affects all other members and the family unit as a whole. The degree of cohesion, communication and roles and rules are the important variables. They argued that a person's satisfaction in a relationship depends on the quality of their role construction, which (in turn) depends on the clarity and consensus in defining role expectations. For example, if a daughter-in-law believes that she should be responsible for only half of the housework, her construction of her role carries with it implications that her mother-in-law should be willing to be responsible for the remaining half of the housework. This may be in conflict with the understanding of the mother-in-law!

Family theorists who use a social systems approach to understand families under stress view them as a living organism with both symbolic and real structures. These experts study the sub-systems and supra-systems and their interaction to understand how family tries to maintain a steady state. These theories take into account the interaction and communication of the individual members of the family along with the cultural, historical, economic and developmental influences.

The family's relationship with its environment, which includes physical realities, familial, and social support systems in terms of the extent of use made of and support received from familial and social networks, is important to understand the strengths and coping abilities of the family in the face of stress. The specialised roles of family members change with changing needs of the members. This approach would be useful in understanding the impact of institutional care on children and their understanding and approach to care for their parents and grandparents.

Knowledge of how the family is functioning at the current family life cycle stage and the extent to which it has been able to accommodate the transitional will help in understanding the family's strengths, limitations and coping abilities. Such an assessment will also enable to understand whether the current problem is related to the normal stresses of adjustment to the life cycle stage or whether it is related to other factors. The central idea of this perspective is that the family has a life cycle similar to that of an individual and that families pass through distinct stages marked by significant events. These stages are distinguishable by differences in family structure and functioning. But widespread deviations of such norms can bring about new norms, thus leading to family change. Besides, individuals and families who deviate from such family norms may be doing so, as they are trying to conform to the norms imposed by larger changes in some other institution. Such conflicts lead to social change. Besides, developmental theory can make policymakers more sensitive to the needs of families at different stages of development such as families with young children or elderly couples.

Need for Special Care in Old Age

The crux of the matter is that care of older persons particularly those in the age group of 80 years and above and those with disability or severe disability pose serious questions for the informal caregivers. The major issues that confront the caregivers are employment and consequent commitment of women (traditional caregivers); small family size; migration of adult children to different cities and countries; stress on self-achievements; and increased value of money due to consumerist pressures. These demands have to be balanced with the increasing age of the older persons, increased morbidity and consequent demand for care.

There are few related questions that need to be answered by the individual, family, community, government and the private organisations before any satisfactory answers are found. Some of the questions are as follows: Can some of assistive devices be developed at affordable cost and with reasonable availability for older persons which reduce their dependence on others for most of the activities of daily living (ADL) and instrumental activities of daily living (IADL); development of care facilities in the community for older persons especially those with diseases like dementia; availability of trained human power for home care? In other words, can we make sincere efforts to facilitate the informal caregivers, so that they do not feel 'burn-out'?

It is important to understand that in the last four decades, the average life expectancy at birth and at 60 years has increased considerably. Life expectancy at 60 years increased from 10.7 years to 17 years. Another significant fact is the increase in the number of older person over the age of 80 years. In the year 2015, 14% of the elderly population was in this age category and it is expected to be more than 20% of the total elderly population in the year 2050 (United Nations 2015). It is also important to examine the data on marital status and living arrangement of the older person in the country to understand the changes in the informal care system. It is likely that most of the caregivers of these oldest-old would themselves be young-old.

The data on general health and disability of the older person in the country should also be taken into consideration as it will help in assessing the care needs of older persons in the country. According to census conducted in the year 2001, 5% of the older persons suffer from some form of disability. Among the disabled elderly, 56.7% were married. This proportion was uniform across the type of disability. Almost two out of every five disabled aged were widowed; the proportion of widowed females is significantly higher than that of males.

Almost 50% of the adult disease burden in the high mortality regions of the world, which include India, is due to non-communicable diseases. The diseases that afflict older persons most often are cardiovascular diseases, cancer, osteoporosis and/or arthritis, hearing, vision and cognitive impairment and dementia. The older population is at greater risk due to insufficient intake of fruits and vegetables and inadequate regular physical activity which has a significant protective effect against ischaemic heart diseases, stroke, diabetes mellitus, cancer, reducing blood pressure and raising HDL cholesterol levels, in addition to reducing minor anxiety, depression and weight.

The WHO (2006) study, conducted in India in the year 2003, established beyond doubt the growing need for care of older person with increasing age; 29% of older persons in the age group of 80+ faced extreme and 24% faced severe difficulty in performing household chores; 50% had extreme difficulty in vigorous activity and 19% faced extreme difficulty even in moving about, 16% had extreme body pains and aches and 23% had extreme difficulty in self-care and 25% had severe difficulty in self-care. An almost similar percentage of older persons in age group of 80+ faced extreme and severe cognitive difficulty, respectively.

All older persons, particularly those in the age group of 80 and above and those with disability, require special care. It is not easy to define and understand the dimensions of care. It means concern, support and art of nurturing. In modern day societies, care especially for older persons has assumed different dimension because of increasing longevity, increasing number of older persons, changes in social roles, changes in disease pattern and most importantly, changes in the nature and structure of family and community and the socio-economic settings. More and more people are living longer and require care in old age. The family is getting smaller and moving from the realm of romanticised communitarian values to individualistic values.

According to the NSSO (2006) 60th Round Report, among the elderly who lived alone or with their spouse, the location of residence of the family members was 18.5% within the same building; 36.9% within the same village/town; 29.4% outside village or town. If these trends are any indication, then policymakers should contemplate new structures of care for older persons or provide support to the traditional caregivers so that they could take care of older person without disturbing or with minimal disturbance in their own lives.

Type of Care Required by Older Persons

In this background of change, it is important that all stakeholders understand and address the increasing need for care in old age and find new practical and appropriate means to take care of older persons. It could mean partnership among various actors beginning from the individual to family and community to government and private sector. Therefore, it is crucial that facilities for care develop in this country for the older persons. It could range from respite care for the caregivers, convalescent care, long-term care, rehabilitative care, care services at home to case care. Before we debate the factors for the sustainable and meaningful care services for older persons, it is important to understand the type of care services required by older persons.

The term **convalescent** is defined as a person recovering from illness, which is crucial in the context of inadequate or unaffordable nursing care in hospitals.

Rehabilitation is about restoring an individual to fullest capacity, physically, mentally and socially. It requires a team of people like physiotherapist, occupational and speech therapist, nurses to work with the family members to promote recovery. Unfortunately, this is an area of care that our current health and social care system does not address adequately. Therefore, the main sources of support for the older person in these circumstances are likely to be a family member, neighbour, friend or relative or a voluntary agency.

Respite care is defined as temporary relief. It may be provided for the older person or the carer and may take the following forms: short break for the older person (holiday); increased support to allow the carer to pursue his/her own interests; short stay in a respite care home to enable the carer to take a longer break; day care facilities.

Long-term care includes activities undertaken for people requiring care by informal caregivers (family, friends and neighbours), by formal caregivers, including professionals and auxiliaries (health, social and other workers), and by traditional caregivers and volunteers (WHO 2000). The need for long-term care is influenced by changing physical, mental and/or cognitive functional capacities that are in turn, over the course of an individual's life, influenced by the environment and, therefore, often difficult to predict. The goal of long-term care is to ensure that an individual who is not fully capable of long-term self-care can maintain the best possible quality of life. It could be home-based or institutional.

Home-based care is already prevalent in all developing countries and now also expanding in developed countries. In India, due to inadequacy of medical facilities, most ailments are treated in the home and most people die at home.

These developments require the adoption of a very different approach to health sector policy and health care services: a disease-specific approach alone would no longer be appropriate. Moreover, family and other informal caregivers need information, support and skills if they have to provide meaningful care to the older relative.

These are some of the care models that have been evolved in societies to deal with the issues of care for the increasing number of older persons. These services are provided by the support of the local and national governments, family, local community, NGOs and volunteers. In the developed countries, most of these services are in the organised sector, with government providing necessary financial and regulatory support; however, in the developing countries, the health policies and infrastructure are geared towards only curative services for various diseases. So, the informal support is critical element in age care. Therefore, it is time to make paradigmatic shift to an integrated approach to health which includes care.

Existing Old Age Homes in India: Some Critical Issues

Various empirical studies on old age homes (OAHs) conducted in Kerala, Andhra Pradesh, Tamil Nadu, Maharashtra, Punjab, Delhi and Gujarat reveal the following trends. Most of the OAHs were either free (set up by respective governments, philanthropic organisations) or 'pay and stay' type set up by completely unregulated private operators. Most of the homes were set up as a response to the locally felt need and not a part of a well-thought-out plan. The facilities varied widely. In some homes, residents were expected to pay more compared to the facilities offered, and in some others, small payment entitled them to better facilities. None of the homes had permanent medical facility. The arrangements ranged from daily to weekly visits by doctor; but in all old age homes, the cost of medicine was wholly borne by the resident. There was no provision for the very sick and the bed-ridden older persons. Common health problems reported by the residents included blood pressure, rheumatism and joint pains, diabetes, asthma, heart disease, hearing problem, paralysis.

The residents in 'free' old age homes were mainly very poor and destitute. In some homes, the residents were getting benefit of monthly social pension under the

Indira Gandhi National Old Age Pension Scheme (IGNOAPS). In the 'paid' old age homes, either the residents paid or their children or relatives paid the bill. In many 'paid' OAHs, the residents were expected to deposit a lump sum at the time of admission primarily to take care of any emergency situation later in life.

Most of the residents in old age homes were in the below 80-year category, mostly widows/widowers, those who never married, people with low and medium educational qualifications and hence lower earning ability. Most of them did not work, not many contributed to the chores in the old age home.

Most of the residents were living in the homes out of compulsion and would rather go back to the family. The major reason for living in OAHs was lack of resources and domestic conflict and dearth of caregivers in the family. The reasons could be empirically traced to different segments. For example, the poor had no resources or resources were snatched away by children and had no caregivers; the economically better-off had domestic conflict; males cited practical difficulties and females cited emotional reasons for moving to an institution. Very few residents cited desire for independent living as the reason for moving into an old age home.

In charity old age homes, literacy levels were low. Elderly women were less educated than men. Relatively better educated, higher caste people lived mostly in 'paid' OAHs. Only those who were defrauded by their children lived in free facilities. The number of children was inversely proportional to the transfer into the OAHs; i.e., lesser the number of children, more the chances of moving into the old age home. Most of the residents had lived in extended families before moving into the OAHs and many wanted to go back. But in OAHs where people had lived for longer duration, they were happy to be in the institution, as they had developed their own social network there.

The daily routine in the homes was almost similar: wake up in the morning at about 6.30 am drink tea, do activities of daily living, eat breakfast, read newspaper, in-house socialising, eat lunch, afternoon nap, evening tea, some activity, television viewing, dinner and to bed at 10.00 pm. Many suffered boredom, loneliness and isolation due to limited mobility. The residents' contact with the outside world was much less compared to those who lived in families. They did not often go out to the market, or to meet relatives and friends. The entertainment facilities were limited to television, newspaper, magazines and indoor games. Men preferred reading newspapers and women watching television. The residents, irrespective of the type of OAHs, were generally satisfied with the services. But many of those in the middle income and moderate 'pay and stay' OAHs showed dissatisfaction with the quality of services. It is necessary that some of the basic facilities like toilets, drinking water and food should be given extra consideration. Moreover, the regional character of the residents should be taken into account while planning and preparing food in OAHs.

Recreation is a very important aspect of life. It enriches the otherwise dreary and boring life, as elderly get accustomed to their routine. The OAHs left very little space for recreation, which led to a vicious circle of non-participation, boredom and pessimism. The data also showed that quite a significant percentage of elderly were uncertain and worried about their future, so they felt nervous and concerned. Their worries were due to their lonely existence (no one to share their distress or soothe them in difficult times). Their solitary existence was responsible for their unhappiness even after shifting into the OAHs.

According to the data available with HelpAge India in the year 2009, there were 1280 OAHs in India and 50% of these OAHs were in southern states, 18% in eastern region and 18% in western region (HelpAge India 2009). The lowest number of OAHs was in the northern region. Of OAHs, 43% provided residential facilities for both men and women, 10% catered to 'only women' and 4% were for 'men only'. The total number of seats in the listed OAHs was approximately 44,000. Of these seats, 51% were in OAHs in the southern states, 20% in western, 18% in northern and 10% in eastern region; 42% were providing free facilities to the residents and 20% were 'paid' OAHs; 13% provided both free and paid services. Only 42% had medical aid available for the residents. Of these organisations, 55% were registered under the Societies Registration Act 1860.

Way Forward: Need for Some New Initiatives

OAHs are not only a response of the increasing need to house but also care for the burgeoning population of the older persons in the country; though, a comprehensive informal local care system would not only be most appropriate for the older persons, but also be functionally efficient and cost-effective. In the context of far-reaching changes in the economy and concomitant changes in the culture, it is imperative for us to look to doing things 'differently'. Family and community could still be the mainstay of care for the older persons, but some rudimentary changes are essential for a viable system for the future.

The family efforts could be supplemented by the volunteers or paid local community workers who would be in regular contact with the families. These workers would undertake a number of health promotion, curative and community development activities and receive more practical training than any trained community volunteers.

These suggestions for local informal structures of care, however, should not be beyond the purview of tests for quality assurance. There should be a system for training and supervision of formal and informal caregivers, information system development, standard setting and development of basic guidelines. In addition, legislation can play a part in promoting quality. An important element in the quality of home-based care is the capacity of families as caregivers; this can be enhanced by social recognition, training, counselling and appropriate support (respite services, assistance, material support, etc.).

Traditionally, there was an inherent principle of collective ownership in any property acquired in and by the family. It belonged to everyone in the family, and it was based on respect for the elderly with a right to occupy a portion of that property according to their need. But this has been replaced by the concept of private property and individual ownership. The traditional ethos of parents saving enough to give to the next generation has been replaced to a large extent by individual consumption at the level of each generation. The following generations in every family or clan grew on the repatriated/saved resources of the elderly; but this interdependence had now been replaced by self-sufficiency and independence to a large extent.

As stated above, the need for institutional care for older persons is being increasingly felt. To provide effective care to older persons in future, we need to understand the various needs and respond to these specific needs. There will be the destitute and/or poor elderly who will require special OAH as they will lack resources, both human and material. There will be the able-bodied older persons who may require home care based on their level of independence and need for emotional bonding. Then, there will be those requiring special nursing homes. There are disabled, frail, dependent elderly who will require special assistance homes to take care of their special needs. As of now, there is a yawning gap in all these services. Despite the provisions of NPOP and MWPSC Act, nothing substantial has been done to address these needs. OAHs are still primarily run by charity and private organisation without any guidelines, specifications and regulation. They are addressing the housing and care needs of only a fraction of the population, and for that too, we have no way of knowing the quality and appropriateness of the services. OAHs are still the last resort so to say for the elderly and are mainly providing very basic services for the residents.

Caring for an older relative or friend can be physically and emotionally demanding, especially for individuals who combine work and care for other family members. It could be worse in case of the caregiver being alone and old himself/herself. The trends of living arrangements mentioned above point to the fact that in the coming decades, in India, many older persons will be without the complete support of their young adult children and will probably have to depend on the spouse or siblings for care. Therefore, care system should be devised to reduce the psychological and emotional stresses they are likely to face. At the same time, the care system should provide a positive experience for the older person as well.

The health care system should develop capacity to take care of the convalescent and rehabilitative care needs of older persons. This would require specialised intervention and may require more than one expert. It is important to mention in this context that according to WHO (2006), per 100,000 population in India following facilities were available in the year 2002: 59 allopathic doctors; about one hospital and 67 beds; aggregate number was 89 (including dispensaries, CHCs, PHCs, sub-centres, sanatorium and TB clinics and other health establishments). Given these limitations, the policymakers would have to think ‘out of the box’ to integrate this system with provision of care.

Respite care services should be available in all cities and towns, which could be in the form of day care centres, short stay homes, home-based care services to provide quality care to older person and relief to the caregiver. These services could be very meaningful for the older persons especially those suffering from cancer and dementia, and their families. It would allow older person to remain in their homes in the company of their loved ones without overburdening them. Long-term care services may be able to provide benefit to the frail older persons requiring complete care, and those suffering from multiple disabilities may benefit the most out of such services.

There are a couple of important questions in provision of care services to the older persons that the policymakers should grapple with: development of infrastructure, management of care services, financial viability, monitoring and control.

The first step in achieving integrated health care system would be coordination of actions at all levels: various levels of government, in the private and not-profit, voluntary sector. It is also essential to ensure that home-based care is an integral part of overall primary health care and health system reforms. Indeed, such reform provides an excellent opportunity to develop and implement, or strengthen, home-based long-term care. The communities would have to be prepared for such a role. The National Rural Health Mission should have elements on long-term home care system to provide relief to the poor older persons in rural areas, who are anyway on the periphery of the existing health system. Some other major steps could be including and encouraging grass root initiatives; integrating it with primary health care; using local structures; including family and community in decisions about health services and the provision of care; in-built mechanisms to deal with changing needs of the population; including the concerns of the most marginalised segments of the older population like migrant workers, those displaced by civil unrest and by natural and other disasters; in-built mechanisms to learn from good and bad practices of others; structural, planning, budgetary and functional integration of the care system into the health and social system.

There is also an urgent need for the academic bodies to be concerned about this segment of the population and encourage research scholars to take up studies on the important aspects of lives of older persons. If OAHs are part of the system that will respond to the needs of elderly, then a few steps are essential and these are as follows: government to make general guidelines for setting up and running OAHs in the country; sensitise the architects and builders to develop elder friendly designs and buildings, pool of professionally qualified staff to manage the OAHs and provide care and specialised institutes like in the hospitality industry to develop the skills. It is important for the government to facilitate development of OAHs that are capable of providing shelter, care and responding to the emotional and self-fulfilment needs of the older persons.

References

HelpAge India. (2009). *Directory of old age homes in India*. New Delhi: HelpAge India. http://socialjustice.nic.in/writereaddata/UploadFile/Annexure-VIII.pdf.

National sample Survey Organisation. (2006). *Ministry of statistics and programme implementation, government of India morbidity, health care and condition of the aged*. New Delhi: Govt of India.

United Nations. (2015). *World population ageing*. New York: United Nations.

World Health Organisation. (2000). *Home-based long-term care; WHO technical report series 898*. Geneva: WHO.

World Health Organisation. (2006). *Health system performance assessment; World health survey 2003*. New Delhi, New Delhi: IIPS, Mumbai & WHO Geneva and WHO India WR Office.

Chapter 6
Elder Care and Living Arrangement in Kerala

S. Irudaya Rajan, S. Sunitha and U.R. Arya

Abstract Kerala is an ageing society wherein population ageing is proceeding at a very fast pace. Given the context of ageing in Kerala, the living arrangements of the elderly have an important role, particularly when it comes to elder care. Using the Kerala Aging Survey 2013, this paper evaluates the scenario of family care for the elderly, the perception of elderly about their care, association between health patterns and living arrangements and the determinants of living arrangements of the elderly. Bivariate analysis and multivariate logistic regression were used for the analysis. The findings of the study revealed that living arrangements of the elderly play an important role in elderly care. Among those who live with partners, more than 45% were of the view that it is the duty of children to take care of their parents in old age. More than 43% of the elderly who live with their partner consider living with their partner as the best option. Among those who live with married sons, nearly 60% perceived that such a life is best for them. Among those who stayed with sons and daughters, the share of those who need special care is comparatively greater. Living arrangements determine the source of care provision when special care is required. Results of the regression analysis showed that compared to female elderly, male elderly are less likely to live alone. Further, it was found that compared to female elderly, male elderly are more likely to live with a daughter.

S. Irudaya Rajan (✉) · S. Sunitha · U.R. Arya
Centre for Development Studies, Prasanth Nagar, Medical College P.O, Ulloor, Thiruvananthapuram, Kerala 695 011, India
e-mail: rajan@cds.ac.in; rajancds@gmail.com

S. Sunitha
e-mail: sunithasyam@gmail.com

U.R. Arya
e-mail: aryaur1111@gmail.com

S. Irudaya Rajan and G. Balagopal (eds.), *Elderly Care in India*,
DOI 10.1007/978-981-10-3439-8_6

Introduction

Increase in life expectancy coupled with reduction in birth rate leads to ageing of population. "Ageing of population" is defined in terms of the proportion of persons aged 60 years and over in the total population. As per the United Nations World Population Projections to 2150, the percentage of older population in the world population is expected to increase from 9.5% in 1995 to 30.5% in 2150. In India, the size of the elderly population is growing fast, although it constituted only 7.4% of total population at the turn of the new millennium (Situation Analysis of the Elderly in India 2011). Of the total population of Kerala, the proportion of aged population increased from 10.5% in 2001 to 12.6% in 2011 and the proportion of 80+ population in the total population increased from 11.7% in 2001 to 12.9% in 2011 (Census 2011). The pace of population ageing in Kerala is very fast. Kerala is thus an ageing society. Kerala is known for its high levels of out-migration, a factor that might have contributed to the high proportion of the elderly (Gulati and Rajan 1999).

In the Indian scenario, family care is an integral part of old age care (Government of Kerala 2009). Given the context of ageing in Kerala, living arrangements of the elderly have an important role to play in their care. The terms "living arrangement" or "co-residential arrangement" of the elderly refers to the household structure of the elderly (Palloni 2001). In particular in this era of increasing longevity, there is a likelihood of a rise in the cost of providing care for the elderly. The pattern of living arrangements of the elderly varies from country to country (Kamo and Min Zhou 1994). Living alone or living with spouse is the most common form of living arrangement found in the developed countries and residing with children is the most common pattern in the developing countries (United Nations 2005). This pattern of living arrangement is mutually beneficial to both children and elderly (Beard and Kunharibowo 2000). It is a common assumption that co-residence with children and grandchildren in multigenerational households benefits the elderly and elderly who live with at least one child are better off and better provided for than those who live alone or with non-relatives (Teng 2007).

In recent times, there has been a progressive decline in the support for elderly (Irudaya Rajan and Mishra 1997). As per the Kerala Ageing Survey 2013, only 36.44% of the elderly receive pension from the government. It is also well known that elderly people belong to an age group, which demands not only financial support, but also emotional and physical assistance to help them lead a dignified life. Happy are the ones who could avoid disease and disability. However, this group is just a small minority. As one approaches old age, chances of developing some kind of disease or disability are high. In this context, the role played by family is immense. Family is the most cherished social institution in India and the most vital non-formal social security for the elderly (Central Statistical Organisation 2006). In recent years, many changes have occurred in our family system. While there has been an increase in the proportion of the elderly living with their spouse only, it could be due to increase in male life expectancy. The erosion of the traditional joint family system is mainly due to modernisation and urbanisation. Shift from joint family system to nuclear families has an impact on every member in the household or family. This impact on the elderly is very adverse and powerful, in most of the cases. Now with children flying the coop to where the money and opportunities beckon, the elders are left alone wrestling with functional disabilities, loneliness, and sometimes with no money (The Hindu 2011). The most crucial aspect of living arrangements for the elderly is co-residence with adult children in extended families or multigenerational households where kin provide income, personal care, and emotional support to the elderly (Irudaya Rajan and Kumar 2003).

To a large extent, cultural variations also affect the living arrangements of the elderly with sons or daughters, and whether they will stay with the elder son or younger son (Goode 1963). The living arrangement is affected by the perception of cost and benefit by both older and younger generations manifested in preferences and constraints (United Nations 2001).

The other side of the coin is looking at elderly care from the perspective of the caregivers in family. It is often seen that a healthy and earning elderly member is always considered an asset. Older adult who work would be welcomed with open arms. On the other hand, bed-ridden and fully dependent parents may be as a liability even by their own children. The elderly, who have adequate financial resources and take care of family affairs, who are productive and involved in income generation activities and contribute to the family, are looked after well and held in good esteem by the young.

Still, a higher proportion of the population aged 60 and over lives with their family members in Kerala. This does not mean that the state does not have people who live alone. Compared to people who live alone, elderly people who stay with their family members are supposed to benefit from more care and attention. Elderly people who live with their family members do not feel isolated. Living with family members not only increases social interaction, but also strengthens emotional connectedness. It is very essential that elderly people who live separately from their children be given some kind of care and support. This gives rise to the question as to whether children are staying far away from or close to where their elderly parents stay. If children stay near, it would be possible for them to visit their parents more often and to provide the care and support they need. In many cases, the elderly may choose to live alone. Though they wish to enjoy life with their children and grandchildren, they are left alone and forced into an isolated existence. An elderly person who has lost a spouse and lives with the children has to make a lot of adjustments in life compared to those who live with their spouse.

Co-residence is a way of minimising cost also. It avoids cost of constructing an additional building. In the context of high house rents in urban areas, living with one's old parents has multiple benefits. However, in many cases, weakening of inter-generational relationships creates problems between aged parents and children. Such problems may lead both elderly parents and their children to live a life of separation and isolation. Thus, in order to provide care to the elderly group of population, it is important to mutual understanding between the older and younger generation. The kind of support and care experienced by the elderly depends on the living arrangements of elderly, though their own children extend such type of support, in many cases.

The current social, economic, and related factors will have profound influence on the future of the elderly. Therefore, it is important to understand the differences between care and support for elderly people. “Support for the elderly” and “taking care of the elderly” are two different concepts. While support for the elderly is defined as providing financial assistance (pensions and social security), care of the elderly is defined as extending emotional support, which can be provided only by family members or by those persons with whom the elderly live (Irudaya Rajan and Kumar 2003).

Under the Maintenance and Welfare of Parents and Senior Citizens Act 2007, a senior citizen, including parents, who is unable to provide maintenance for himself/herself from his/her own earning or out of the property owned by him/her shall be entitled to get support from his/her children because, as per this Act, the obligation of the children to maintain his or her parent extends to the needs of such parent either father or mother or both, as the case may be, so that such a parent may lead a normal life. Though the elderly are provided with care by their own family members, there may be a mismatch between what is expected by the elderly and the care or support they actually receive. The nature of this varies from monetary expenditure to time spent by the family members in taking care of the elderly. Whether the older generation is satisfied with the care given to them by the younger generation is still an unanswered question. However, if an older person is healthier than his peer group, he will have not only a healthy body, but also a healthy mind.

It is also very important to look at the type of care when it comes to the difference between disease and disability associated with the old age. The type of care required for disease and disability is different. Furthermore, care demands differ from one type of disease to another and one type of disability to another. In old age, a person with a disability may like to stay with his/her children. Whether this dream is realised or not depends on many factors such as place of residence of children and the nature of their employment. The mental health of an elderly person is equally important in this context. An elderly, disabled person affected by some kind of chronic disease suffers multiple burdens. Locomotors disability is usually associated with the oldest-old. Chronic diseases also attack the oldest-old very easily. This brings up the question of income and emotional support to provide for care. Who is it that provides care for the elderly during times of disease and disability? What are the perceptions of the elderly regarding elderly care? What determines the living arrangements of elderly? Is there any difference in family care to elderly? This paper attempts to look at these questions.

Objectives

- To find out the differences in family care for the elderly
- To understand the perceptions that the elderly have about elderly care

- To examine the association between health patterns of the elderly and their living arrangements
- To understand the determinants of the living arrangements of the elderly

Area of Study

In this chapter, we discuss the living arrangements of the elderly with respect to elder care. There are several living patterns for the elderly such as living alone, living with spouse, living with children and living in old age homes. Living alone or with spouse is the most stable living arrangement for people who are young–old, whereas, for the oldest–old, living with a child or grandchild is the most stable arrangement (Wilmoth 1998). However, whenever they need care, it is important to know who the main providers of care are. Though the son takes care of the elderly, the main caregivers are the female members of the family, especially the daughters-in-law. In this context, this chapter describes the satisfaction with care received by the elderly by their living arrangements.

Data and Methodology

In order to highlight the plight of the elderly in Kerala, the Centre for Development Studies, Kerala, conducted a survey entitled “Kerala Aging Survey (KAS) 2013”, which covered 14 districts of Kerala. The 2011 Census data on aged population in Kerala were used for the sampling purposes. The samples of KAS 2013 were obtained from Kerala Migration Survey (KMS) 2011. Those households which have people above 60 years of age and those which had people who would attain 60 years within the next 2 years were taken from KMS 2011 to form samples of KAS 2013. There were 15,000 households in the Kerala Migration Survey 2011 out of which 5,713 households had elderly people. KAS 2013 covered 7,768 elderly people. Though there were 7768 samples, only 6702 were selected for the analysis. The selection was made by considering four categories of living arrangements such as living alone, with spouse, with married son and with married daughters, as these were the major types of living arrangements in the state.

The methodology followed involves bivariate analysis and multinomial logistic regression. For our analysis, we categorised the elderly into three age groups: 60–69 (young–old), 70–79 (old–old) and 80+ (oldest–old). In bivariate analysis, the association between the living arrangements of the elderly and other variables such as district, place of residence, gender, age group, religion, marital status, caste, disabilities, chronic disease, main caregiver, daily meals and satisfaction regarding care were taken into account. Further, living arrangement is analysed based on the perceptions of the elderly about elderly care. To identify the determinants of living arrangements of elderly, a multinomial logistic regression analysis was carried out by fixing the four categories of living arrangements mentioned above as the dependent variables and variables such as age, sex, place of residence, religion, depression level, caregivers and satisfaction of care as the independent variables.

A linear regression model is used when the dependent variable is nominal with more than three values, namely tetrachotomous. The dependent variable is tetrachotomous and the independent variable is nominal or continuous. It is used to estimate the degree of effect of each independent variable on the dependent variable. It compares multiple groups through a combination of binary logistic regressions. The comparison of the dummy-coded dependent variable done with the highest numeric score is used as the reference group. It does not make any assumptions of normality, linearity and homogeneity of variance for the independent variables (Garson 2008).

The Model

The multinomial logit model is based on the assumption that the log odds of each responds follow a linear model.

$$\log\left(\frac{\pi_i}{\pi_j}\right) = \alpha_j + \beta_j X$$

where α_j is a constant and β_j is a vector of regression coefficients, for j = 1, 2, ..., j − 1. This model is analogous to a logistic regression model, except that the probability distribution of the responds is multinomial instead of binomial and j − 1 equations instead of one. The j − 1 multinomial logit equations contrast each of the categories 1, 2,..., j − 1 with category j, whereas the single logistic regression equation is a contrast between success and failures. If j = 2, the multinomial logit model reduces to the usual logistic regression model (Hosmer and Lemeshow 1998).

Significance Test of the Model Log Likelihood

Likelihood is a probability, which varies from 0 to 1. The log likelihood (LL) is its log and varies from 0 to minus infinity (it is negative because the log of any number less than one is negative). Log likelihood is the basis for tests of a logistic model (Garson 2008). The initial log likelihood function (−2 log likelihood or −2 LL) is a statistical measure like total sums of squares in regression. If the independent variables have a relationship with the dependent variables, it will improve the ability to predict the dependent variable accurately and the log likelihood measure will decrease. The initial log likelihood value is a measure of a model with no independent variables, that is, only a constant or intercept. The final log likelihood value is the measure computed after all of the independent variables have been entered in the logistic regression. The difference between these two measures is the model chi-square value that is tested for statistical significance (Kleinbaum 1994).

Identifying the Statistically Significant Predictor Variables

There are two outputs related to the statistical significance of individual predictor variables: Likelihood Ratio Tests and Parameter Estimates. The likelihood ratio tests indicate the contribution of the variable to the overall relationship between the dependent variable and the individual independent variables. The parameter estimates focus on the role of each independent variable in differentiating between the groups specified by the dependent variable (Kleinbaum 1994).

Odds Ratios with 95% Confidence Interval

The log odds ratio is the ratio of two odds and it is a summary measure of the relationship between variables. It gives the relative amount by which the odds of the outcome increase (odds ratio greater than 1) or decrease (odds ratio less than 1), when the value of the independent variable is increased by 1 unit (Kleinbaum 1994). The dependent variable "living arrangement of elderly" is a tetrachotomous dummy variable in which living alone is considered as the reference category.

Socio-Demographic Characteristics of the Elderly

In our selected sample, the distribution of the elderly across the state shows that though the share of elderly is the highest in Kozhikode district, the proportion of the 70–79 age group is higher in Palakkad district and Ernakulam district tops in the proportion of elderly who fall under the 80+ age group. Here, the share of elderly who belong to the 60–69 age category is higher compared to those in the 70–79 and 80+ age groups. Of the total, 53.8% of the elderly in Kerala belong to the 60–69 age group, 32.5% belongs to the 70–79 age group and only 13.7% falls under the oldest-old category (Table 6.1). Though in all these three age groups, the share of the 80+ category is comparatively less, and it is found that advancements in medical science have contributed towards increasing longevity.

The proportion of aged women in Kerala was 4% higher than that in the whole country (Census 2011). The sex ratio of Kerala is very unique. It was observed that 79% of the elderly lived in the rural areas, while 21% resided in urban areas. The

Table 6.1 Socio-demographic characteristics of the elderly (as %)

Characteristics	Categories	Percentage
Age	60–69	53.8
	70–79	32.5
	80+	13.7
Sex	Male	40.9
	Female	59.1
Place of Residence	Rural	79.0
	Urban	21.0
Marital Status	Unmarried	0.5
	Married	65.0
	Widowed	33.5
	Separated or divorced	1.1
Educational qualification	Illiterate	13.7
	Below primary	16.5
	Primary	14.6
	Upper primary	22.9
	High school	9.4
	Secondary and higher secondary	16.2
	Degree and above	4.4
	Others	2.2
Religion	Hindu	61.1
	Christian	21.2
	Muslim	17.7
Caste	Brahmin	2.1
	Nair	15.4
	Ezhava	24.4
	Viswakarma	2.9
	SC	7.6
	ST	1.4
	Others	46.3
Living arrangement	Living alone	3.0
	With spouse	29.9
	With son	56.3
	With daughters	10.7

Source: Data compiled from Kerala Ageing Survey 2013 undertaken by the Centre for Development Studies, Kerala and sponsored by Government of Kerala

case of Pathanamthitta district needs to be highlighted here because nearly 97% of the elderly lived in rural areas compared to a small minority who lived in urban areas.

The survey revealed that 59.1% were females. This shows the female-dominated picture of elderly people in Kerala and goes to prove that sex ratio is in favour of females among both children and elderly females in the state.

Majority of the elderly are married. About 34% are widowed. Looking at their educational qualification, about 23% had an upper primary level of education. Nearly 17% of the elderly had below primary level of education. However, almost a similar share of elderly had secondary level of education. Even though Kerala is the state with the highest level of literacy in India, nearly 14% of the elderly are illiterate. Religion-wise and caste-wise distribution of the elderly is similar to the general trend in Kerala. Living alone is a rare case in Kerala. More than half of the elderly are living with a son. Nearly 30% are living with their spouse.

Results

Difference in Family Care for the Elderly

This section looks at the association between living arrangement and various variables such as district, place of residence, gender, age, religion, caste and marital status.

Table 6.2 shows the distribution of the elderly in each district along with their living arrangement. As per the figures, 56.3% of elderly live with their son. The share of elderly living with son is high in all the districts and less than 30% of the elderly living with spouse, and the share of elderly living with their daughters is

Table 6.2 Living arrangement of elderly by district

District	Living with partner	Living with son	Living alone	Living with daughters	Total
Thiruvananthapuram	38.7	45.5	3.6	12.1	100.0
Kollam	29.8	56.9	4.0	9.3	100.0
Pathanamthitta	39.4	47.5	2.3	10.9	100.0
Alappuzha	21.6	60.1	5.1	13.2	100.0
Kottayam	28.8	59.9	1.2	10.0	100.0
Idukki	31.8	59.0	3.3	5.9	100.0
Ernakulam	32.0	56.7	2.5	8.7	100.0
Thrissur	32.6	57.0	2.2	8.2	100.0
Palakkad	25.0	57.3	4.0	13.7	100.0
Malappuram	13.3	76.5	1.5	8.8	100.0
Kozhikode	39.1	49.7	4.3	6.8	100.0
Wayanad	31.0	60.0	1.7	7.3	100.0
Kannur	33.9	42.6	3.7	19.8	100.0
Kasaragod	19.4	66.0	1.7	12.9	100.0
Total	29.9	56.3	3.0	10.7	100.0

Source: Data compiled from Kerala Ageing Survey 2013 undertaken by the Centre for Development Studies, Kerala and sponsored by Government of Kerala

Table 6.3 Living arrangements of the elderly by gender

Gender	Living with partner	Living with son	Living alone	Living with daughters	Total
Male	29.8	56.2	1.2	12.8	100.0
Female	30.1	56.4	4.3	9.2	100.0
Total	29.9	56.3	3.0	10.7	100.0

Source: Data compiled from Kerala Ageing Survey 2013 undertaken by the Centre for Development Studies, Kerala and sponsored by Government of Kerala

comparatively less. One notable finding is that 3% of elderly live alone. The share of this category is low in all districts.

Gender-wise analysis reveals that there is no considerable difference between the proportion of male and female elderly who live with their partner and with their sons (Table 6.3). In contrast to this, there is a substantial difference between the share of male and female elderly in the case of those who live alone and living with daughter. In the case of living alone, the share of females is greater, whereas in the case of living with daughter, the share of males is higher.

General Profile of the Elderly by Living Arrangements

Examination of the age-wise data shows that more than 40% of the elderly in the 60–69 age group, 62.2% of elderly in the 70–79 age group and 71% of elderly in the 80–89 age group stay with son. The proportion of elderly living alone is comparatively lesser in all the three age groups.

Religion-wise analysis shows that in all the three religions under consideration, the proportion of the elderly who live with son is higher. Caste-wise distribution among elderly also shows the similar picture. Here also, the proportion of elderly living alone is quite low (Table 6.4).

The role of marital status is very important when it comes to determining whether an elderly person receives adequate care and attention during the twilight years of life. Among divorced, married, separated and widowed, the proportion of elderly living with son is high.

Perception of the Elderly About Their Care

Transfer of Property

Perceptions of elderly regarding transfer of property show that almost 28% of the elderly perceived that it is good to transfer the whole property to their children during their lifetime (Table 6.5). Here, there is no difference among those who stay with son/daughter/spouse/alone. More than one-fourth (28%) of the elderly were of

Table 6.4 Socio-demographic profile of elderly by their living arrangements

	Living with partner	Living with son	Living alone	Living with daughters	Total
60–69	38.1	49.1	2.7	10.1	100.0
70–79	23.0	62.2	3.4	11.4	100.0
80+	14.4	71.0	3.5	11.1	100.0
Religion					
Hindu	30.8	54.3	3.7	11.2	100.0
Christian	34.9	55.0	2.6	7.5	100.0
Muslim	21.0	65.0	1.4	12.5	100.0
Caste					
Brahmin	51.1	31.9	7.1	9.9	100.0
Nair	36.6	46.9	4.1	12.3	100.0
Ezhava	29.4	56.6	3.6	10.4	100.0
Viswakarma	22.3	64.8	1.0	11.9	100.0
SC	24.3	60.4	3.4	12.0	100.0
ST	27.5	52.7	3.3	16.5	100.0
Others	28.5	59.4	2.3	9.8	100.0
Place of Residence					
Rural	28.3	58.0	3.0	10.7	100.0
Urban	36.2	50.1	3.1	10.7	100.0
Marital Status					
Unmarried	41.9	12.9	38.7	6.5	100.0
Married	42.9	47.5	0.7	8.9	100.0
Widowed	4.9	74.1	6.8	14.2	100.0
Separated	13.3	66.7	11.1	8.9	100.0
Divorced	29.6	40.7	18.5	11.1	100.0
Total	29.9	56.3	3.0	10.7	100.0

Source: Data compiled from Kerala Ageing Survey 2013 undertaken by the Centre for Development Studies, Kerala and sponsored by Government of Kerala

Table 6.5 Transfer of property

Transfer of property	Living with partner	Living with son	Living alone	Living with daughters	Total
Transfer whole property to their children within their life	23.3	29.4	26.0	29.7	27.5
Keep one share and transfer rest of the property	28.0	29.3	27.0	23.9	28.3
Keep their whole property and transferred only after their death	58.8	50.1	56.9	52.9	53.2

Source: Data compiled from Kerala Ageing Survey 2013 undertaken by the Centre for Development Studies, Kerala and sponsored by Government of Kerala

the opinion that older people should keep one share of the property with them and transfer the rest to their children during their lifetime. Here also there is not much difference in relation to living arrangement. In contrast to this, 60% of the elderly who reside with their spouse are of the opinion that elderly people should keep their whole property with them and it should be transferred to the children only after their death. This kind of feeling could be due to the absence of financial security, which they have been experiencing after transferring all the shares to their children. People often find excuses for extending financial support even to their own parents once the former are well placed and settled.

It is also seen that most of the elderly people who transfer a larger share of family property to the younger son resist going to live with other children, as they believe that they would be a burden to them since they have not given them more property. These feelings exist even though their other children would not consider their parents a burden. Though they regret their decision to stay with their younger son and daughter-in-law, they try their best to cling to the family home. In most of the cases, these old people do not complain to their other children and suffer in silence because they want all their children to live in harmony forever.

About 53.6% of the elderly who felt that older people should remain active participants in labour market after their retirement, lived with their sons. This can be attributed to many factors such as those who stay with son and his family could have less freedom and financial autonomy when compared to other living arrangements. Here, the share of the elderly who lived alone is considerably less.

Perception on Who Should Take Care of Older People

Perception of older people on who should take care of them shows a heterogeneous picture. About 47% of the elderly were of the opinion that it is the responsibility of children to take care of their parents in their old age (Table 6.6). There is difference among the elderly with regard to this, as according to 36.4% sons should bear the care responsibility, whereas only 4.6% of elderly felt that elder care is the

Table 6.6 Perception on who should take care of older people

Perception on who should take care of older people	Living with partner	Living with son	Living alone	Living with daughters	Total
Aged parents should be independent	17.7	7.7	22.5	9.2	11.3
Son or Daughter	45.3	46.1	46.1	56.6	47.0
Sons	31.0	43.4	22.5	18.9	36.4
Daughters	5.2	2.4	5.9	14.3	4.6
Others (specify)	0.7	0.5	2.9	1.0	0.7
Total	100.0	100.0	100.0	100.0	100.0

Source: Data compiled from Kerala Ageing Survey 2013 undertaken by the Centre for Development Studies, Kerala and sponsored by Government of Kerala

responsibility of daughters. Among those who lived with partners, more than 45% were of the view that it is the duty of children to take care of their parents in their old age. Among those who lived alone, 46% stated that parental care during old age is the responsibility of the children.

Perception on Best Place to Live

Perception of elderly on the best place to live also displays an image of similarity and differences. From Table 6.7, it is clear that more than 43% of the elderly who lived with their partner consider that living with partner is the best idea. Among those who lived with married sons, nearly 60% felt that such a life is best for them. This shows that for these elderly, there is no discrepancy between what they wish for and what they really experience. Among those living alone, while 27.9% preferred to live with their married son more than 24% were happy with their current living arrangement. More than 27% of elderly who lived with daughters consider that living with son would be best for them. It can also be noted that though fewer in number, many people preferred a life in old age homes because they might think that in old age homes they could have more peace of mind and happiness. These feelings may arise out of their experience with other members in the family. It is a natural tendency to have such feelings if one is living alone. However, even when residing with children or partner, if one feels the need to live in old age home, then arise a question of kind of support that person gets from them.

Table 6.7 Perceptions on the best place to live

Perception on best place to live	Living with partner	Living with son	Living alone	Living with daughters	Total
With spouse only	43.7	19.6	18.6	23.5	27.2
With married daughters	21.1	16.7	19.6	43.1	20.9
With married sons	30.6	59.7	27.9	27.1	46.6
With unmarried children	2.5	2.9	2.9	4.5	3.0
With other relatives	1.2	0.3	4.4	0.7	0.8
Alone	0.6	0.5	24.5	0.8	1.3
Old age homes	0.2	0.2	2.0	0.3	0.3
Total	100.0	100.0	100.0	100.0	100.0

Source: Data compiled from Kerala Ageing Survey 2013 undertaken by the Centre for Development Studies, Kerala and sponsored by Government of Kerala

Member in Social and Religious Group

Only 10.6% of the elderly were members of one or the other social or religious group. The share of this category is comparatively higher among those who lived with their spouse. This is mainly because here there were no restrictions created by children to prevent their parents from taking part in these activities. Chances are high that elderly who live with children experience many restrictions because of their age.

Separate Room

Only 26.8% of the elderly had a separate room for themselves at their place of residence. A striking point is that none of the elderly who stayed with their daughter enjoyed a separate room facility. More than 25% of elderly who stayed with their son slept in a separate room. Among those who stayed with their partner, only 29% had separate room. Only 26.4% of elderly who lived alone had a separate room. The irony is that a small proportion of those who lived with a son slept in the kitchen/corridor/hall, etc. They lived in poor or Kutcha house as they have no separate room of their own.

Satisfaction with the Staying Facility

Regarding satisfaction about the staying facility, among the elderly who had no separate room, 51.6% were satisfied with their staying facility (Table 6.8). While 52.4% of the elderly who lived with their son were satisfied with the staying facility, 54.1% who lived with their daughters were satisfied with their staying facility.

Table 6.8 Satisfaction with the stay facility

Satisfaction with the stay facility	Living with son	Living with spouse	Living with daughters	Total
Good	39.3	46.7	33.8	40.4
Satisfied	52.4	48.2	54.1	51.6
Not satisfied	8.3	5.1	12.2	8.0
Total	100.0	100.0	100.0	100.0

Source: Data compiled from Kerala Ageing Survey 2013 undertaken by the Centre for Development Studies, Kerala and sponsored by Government of Kerala

Association Between Health Pattern of Elderly and Their Living Arrangements

Disability

A look at the disability aspect shows that among the elderly who lived with their partner, almost 80% suffered from disability related to vision (Table 6.9). Visual disability was high, as almost 80% of the sample faced this problem. Further, Table 6.9 also reveals that almost 44% of elderly suffer from problems associated with chewing, 23% of elderly suffer from hearing-related problems and 33.8% of elderly have difficulty walking. It is also evident from the table that nearly 30% of the elderly in Kerala were in need of special care. The proportion of the elderly who stayed with son and daughters is comparatively higher. This indicates that old people who were in need of care and help are looked after by their children.

Main Provider of Care

It is evident from Table 6.10 that the living arrangement determines who provides assistance at the time of need of special care. Majority of the elderly who lived with spouse was looked after by their spouse, majority of the elderly who lived with sons were taken care of by their son, and majority of the elderly who lived with daughters were taken care of by their daughters. Either the son or the daughter takes the responsibility of providing care even if a parent is living alone. In other words, even if the old parents do not reside with them, the children extend a helping hand to their older parents in times of need.

Table 6.9 Disability

Disability	Living with partner	Living with son	Living alone	Living with daughters	Total
Vision	79.9	79.2	76.0	79.9	79.4
Chewing	41.4	45.9	40.7	46.3	44.4
Hearing	18.9	25.8	24.0	26.6	23.8
Walking	28.1	36.4	33.8	35.7	33.8
Need special care	21.1	34.1	28.4	32.4	29.9

Source: Data compiled from Kerala Ageing Survey 2013 undertaken by the Centre for Development Studies, Kerala and sponsored by Government of Kerala

Table 6.10 Provider of care

Main provider of care	Living with partner	Living with son	Living alone	Living with daughters	Total
Wife/partner	52.4	20.4	5.2	25.4	27.3
A son or daughter	34.4	67.8	43.1	64.2	59.6
Member of the family	5.2	6.8	10.3	3.0	6.1
A person residing in the household, not a member of the family	1.2	1.1	3.4	2.2	1.3
A person residing in the household who is paid for this	1.9	1.2	10.3	0.9	1.6
A person not residing in the household	0.2	0.2	0.0	0.0	0.1
A person not residing in the household who is paid for this	0.0	0.1	3.4	0.4	0.2
Unknown	0.2	0.0	0.0	0.0	0.0
Nobody	4.2	2.2	24.1	3.9	3.4
Another situation	0.0	0.2	0.0	0.0	0.1
Persons who live in the residence (or assisted living complex) where you reside	0.2	0.2	0.0	0.0	0.1
Total	100.0	100.0	100.0	100.0	100.0

Source: Data compiled from Kerala Ageing Survey 2013 undertaken by the Centre for Development Studies, Kerala and sponsored by Government of Kerala

Table 6.11 Depression scale

Depression scale	Living with partner	Living with son	Living alone	Living with daughters	Total
Low	52.2	53.8	42.2	51.6	52.7
Medium	38.5	36.0	34.8	37.1	36.8
High	9.3	10.2	23.0	11.3	10.5
Total	100.0	100.0	100.0	100.0	100.0

Source: Data compiled from Kerala Ageing Survey 2013 undertaken by the Centre for Development Studies, Kerala and sponsored by Government of Kerala

Depression Scale

Depression scale of elderly in Kerala shows that more than 52.7% of elderly in Kerala has good mental health, as people in this group are not that affected by problems such as depression and related symptoms (Table 6.11). However, 36.8% of elderly in the state face some kind of mental tension. Further, Table 6.11 also reveals that 10.5% of elderly were vulnerable as they confront difficulties due to mental ill health because of depression. It was observed that among the elderly who stayed with their sons, 53.8% do not face as much mental pressure as those who lived alone. Among the elderly who lived with their daughter, 51.6% have stable mental health and among those who lived with their partner the corresponding proportion was 52.2%. The size of such a mentally healthy group is comparatively less among the elderly who live alone. Some kinds of negative feelings are likely to occur when they are confronted with loneliness.

Number of Diseases

Looking at the number of self-reported diseases among the elderly, it was found that 19.1% of elderly stated that they had no disease, whereas 15.3% of elderly reported that they suffered from more than four different types of diseases (Table 6.12). It is clear from table that 21.4% of people who stayed with their partner felt that they were physically fit compared to others.

Chronic Illness

Table 6.13 shows that 29.8% of elderly reported that they did not have any chronic illness, whereas 35.5, 25.3 and 9.4% of the elderly suffered from one, two and more than three kinds of chronic illness, respectively. Among people who lived alone, 33.3% were more confident that they did not suffer from any chronic illness.

Meals Per Day

It is evident from Table 6.14 that 92.1% of elderly had three meals a day. Among those who lived alone, only 87.3% of elderly had three meals a day. Among the elderly who stayed with their partner/son/daughter, more than 90% had meals thrice a day. Also 13.7% of those who lived alone skipped meals. The share of elderly who have only one meal is low in the case of those who lived with their spouse.

Table 6.12 Number of diseases

Number of diseases	Living with partner	Living with son	Living alone	Living with daughters	Total
No disease	21.4	17.9	17.2	19.6	19.1
One	26.6	26.2	32.4	25.9	26.5
Two	22.7	23.6	25.0	23.1	23.3
Three	16.5	15.9	13.2	14.7	15.9
Four or more	12.8	16.5	12.3	16.8	15.3
Total	100.0	100.0	100.0	100.0	100.0

Source: Data compiled from Kerala Ageing Survey 2013 undertaken by the Centre for Development Studies, Kerala and sponsored by Government of Kerala

Table 6.13 Chronic illness

Number of chronic illness	Living with partner	Living with son	Living alone	Living with daughters	Total
No disease	30.7	28.7	33.3	31.9	29.8
One	36.6	35.2	35.8	33.7	35.5
Two	24.3	25.9	23.5	25.3	25.3
More than three	8.5	10.1	7.4	9.1	9.4
Total	100.0	100.0	100.0	100.0	100.0

Source: Data compiled from Kerala Ageing Survey 2013 undertaken by the Centre for Development Studies, Kerala and sponsored by Government of Kerala

Table 6.14 Meals per day

Meals per day	Living with partner	Living with son	Living alone	Living with daughters	Total
One	0.9	1.1	1.5	1.1	1.0
Two	6.4	6.8	11.3	6.9	6.8
Three	92.6	92.1	87.3	92.0	92.1
Total	100.0	100.0	100.0	100.0	100.0

Source: Data compiled from Kerala Ageing Survey 2013 undertaken by the Centre for Development Studies, Kerala and sponsored by Government of Kerala

This shows that during old age, when partners are together without any other person residing with them, they would take the responsibility of taking care of each other.

Multinomial Regression Analysis

The dependent variable considered for the analysis is living arrangement of the elderly in which four categories of living arrangement have been selected. To find the factors affecting living arrangement, multinomial logistic regression is applied by taking the elderly living alone as the reference category. The coefficient under log (P1/P4) represents the effect of predictor variables on living with son (P1) over living alone (P4), log (P2/P4) represents the effect of predictor variables on living with spouse (P2) over P4 and log (P3/P4) represents the effect of predictor variables on living with daughters (P3) over P4.

The output "likelihood ratio test" indicates the overall relationship between the dependant variable and independent variables. The initial log likelihood (LL) value (5063.338) is a measure of a model with no independent variables and the final LL value (4046.159) is the measure computed after all of the independent variables became active. The difference between these two values is the model chi-square value (1017.179). The model fitting information gives the relationship between the dependent variable and combination of independent variables based on the statistical significance. The probability model chi-square was 0.000, less than or equal to

Table 6.15 Model fitting information and likelihood ratio tests

Model fitting information				
Model	Model fitting criteria	Likelihood ratio tests		
	−2 log likelihood	Chi-square	Df	Sig.
Intercept only	5063.338			
Final	4046.159	1017.179	42	0.000
Likelihood ratio tests				
Effect	Model fitting criteria	Likelihood ratio tests		
	−2 log likelihood of reduced model	Chi-square	Df	Sig.
Intercept	4046.159[a]	0.000	0	
Three age group	4335.504	289.345	6	0.000
Sex	4123.179	77.019	3	0.000
Place of residence	4069.638	23.478	3	0.000
Religion	4150.497	104.338	6	0.000
Depression	4061.498	15.339	6	0.018
Taken care	4460.504	414.345	9	0.000
Family care	4099.595	53.436	9	0.000

[a]This reduced model is equivalent to the final model because omitting the effect does not increase the degrees of freedom

the level of significance of 0.01. This indicates the existence of a relationship between the predictor and dependent variables.

The likelihood ratio test explains which variables are statistically significant. It focuses on the role of each of the independent variables on the dependent variable. Table 6.15 demonstrates that all the variables—age, sex, place of residence, religion, depression, taken care and family care—are significant.

The independent variables such as age, place of residence, religion, taken care and family care are statistically significant in the category of elderly living with spouse, whereas elderly with 80+ age, sex, Christian religion, taken care by daughters and sons are statistically significant in the category of elderly living with daughters (Table 6.16). In the category of elderly living alone, sex, Muslim religion, high depression, elderly independency, taken care by daughters and sons, no family care, a little family care are statistically significant.

The older among the elderly are less likely to be living with their spouse as widowhood would occur at some point. As compared to the elderly living in urban areas, elderly in the rural areas are less likely to be living with spouse. Compared to Hindus, Muslims have a less chance and Christians have a higher chance of living with spouse. The elderly living with spouse preferred being independent to being taken care of by their children, either son or daughter. In addition, they opined that daughters should take care of parents. Elderly living with spouse enjoy a great deal of family care.

The oldest-old elderly have less chance of living with their daughter compared to the elderly in the younger age group. Compared to female elderly, male elderly have a higher chance of living with daughter than the elderly does in the younger

Table 6.16 Multinomial logistic regression coefficient for the living arrangement of elderly

		Reference category	Performance (Base = With Son)					
			With spouse		With daughters		Living alone	
			Log (P1/P4)	Exp(B)	Log (P2/P4)	Exp(B)	Log (P3/P4)	Exp(B)
	Intercept		0.097		−1.422		−2.726	
Age	80+	60–69	−1.384	0.251**	−0.315	0.730*	−0.265	0.767
	70–79		−0.768	0.464**	−0.158	0.854	−0.019	0.981
Sex	Male	Female	0.000	1.000	0.374	1.454**	−1.242	0.289**
Place of residence	Rural	Urban	−0.327	0.721**	−0.050	0.951	0.074	1.076
Religion	Muslim	Hindu	−0.579	0.561**	−0.127	0.881	−1.078	0.340**
	Christian		0.191	1.211*	−0.382	0.682**	−0.258	0.772
Depression level	High	Low	−0.062	0.940	−0.012	0.988	0.688	1.990**
	Medium		0.047	1.048	−0.009	0.991	0.012	1.012
Taken care	Aged parents should be independent	Children	0.753	2.122**	−0.044	0.957	1.184	3.269**
	Daughters		0.874	2.396**	1.631	5.109**	1.034	2.811**
	Sons		−0.337	0.714**	−1.024	0.359**	−0.510	0.600*
Family care	None	A great deal	−0.432	0.649**	0.131	1.140	0.998	2.714**
	A little		0.147	1.158	0.246	1.278	1.452	4.270**
	Some		−0.200	0.819*	0.137	1.147	0.114	1.121

*<0.05 and **< 0.001—significant levels

age group. Compared to elderly who belong to the Hindu religion, elderly belonging to the Christian religion are less likely to live with daughters. Elderly living with daughters were of the opinion that daughters should take care of them.

Compared to female elderly, male elderly have a less chance of living alone. Also, compared to elderly who belong to the Hindu religion, elderly belonging to the Muslim religion have less chances of living alone. Elderly living alone have a higher chance of being depressed. The elderly living alone felt that older adults should be independent rather than be taken care of by their children—son or daughter. However, they also opined that daughters rather than son should take care of their parents during their old age. Elderly living alone have less chance of getting family care.

Conclusion

The living arrangements of the elderly play a role in elder care. The share of elderly living with son was high in all the districts. One notable finding is that 3% of the elderly lived alone. As per the study, among the elderly who lived alone, the share of females was more, whereas in the case of living with daughter, the share of males was more. The share of 80+ shows that more than 17% live with their son to get support at their dusk of life. The study also found that 75% of those who lived alone were widowed. Furthermore, 60% of the elderly who resided with their spouse were of the opinion that elderly people should keep their whole property with them and it should be transferred to children only after their death. Among the ones who live with partners, more than 45% were of the view that it is the duty of children to take care of their elderly parents. More than 43% of elderly who lived with their partner consider that living with partner is the best idea. Among those who lived with married sons, nearly 60% perceived that such a life is best for them. Among the ones who stayed with son and daughters, the share of elderly who need special care is comparatively higher. Living arrangements determine who will provide care at the time of need for special care. The size of a mentally healthy group is comparatively less among the ones who stayed alone. Nearly one-fifth of the elderly who stayed with their partner felt that they were physically fit when compared to others. Among people who lived alone, nearly one-third were more confident that they were free from all kinds of chronic illness. Results of regression show that compared to female elderly, male elderly have less chance to live alone. Further, it is found that compared to female elderly, male elderly are more likely to live with their daughter. The elderly living with spouse were of the opinion that the elderly should be independent rather than be taken care of by their children, either son or daughter.

References

Beard & Kunharibowo. (2000). Living arrangements and support relationships among elderly indonesians, DRU.2323, Working Paper Series, RAND, Labour and Population Programme.

Central Statistical Organisation. (2006). *Elderly in india: Profiles and programmes*. Govt of India: Ministry of Statistics and Programme Implementation.

Garson G. David. (2008). Logistic regression. Retrieved from http://wtgrantmixedmethods.com/sites/default/files/literature/Garson_2008_Logistic%20Regression.pdf.

Goode, W. J. (1963). *World revolution and family patterns*. New York: The Free Press.

Government of India. (2011). Census 2011, Office of the Registrar and Census Commissioner.

Government of India. (2011). Situation analysis of the elderly in India, June 2011, Central Statistics Office Ministry of Statistics & Programme Implementation.

Govt. of Kerala. (2009). 'Growing Old in Kerala', UNDP/Planning commission project on strengthening state plans for human development, Thematic study series 6.

Hosmer, D., & Lemeshow, S. (1998). *Applied logistic regression'*. A much-cited recent treatment utilised in SPSS routines. New York: Wiley.

Kamo & Min Zhou. (1994). Living arrangements of elderly Chinese and Japanese in the United States. *Journal of Marriage and Family, 56*(3), 544–558.

Kleinbaum, D. G. (1994). *Logistic regression: A self learning text*. New York: Springer.

Leela Gulati & S. Irudaya Rajan. (1999). The added years: Elderly in India and Kerala, Economic and Political Weekly, *34*(4), WS46-WS51.

Ministry of law and justice. (2007). The maintenance and welfare of parents and senior citizens act, 2007 No. 56 of 2007.

Palloni, A. (2001). Living arrangement of older persons, United Nations Population Bulletin, Special Issue, Nos 42/43, Department of Economic and Social Affairs, Population Division, New York.

Irudaya Rajan, S., & Kumar, S. (2003). Living arrangement among Indian elderly: Evidence from National Family health Survey. *Economic and Political Weekly, 38*(1), 75–80.

Irudaya Rajan, S., & Mishra, U. S. (1997). Population ageing: Causes and consequences in Kerala's demographic transition: Determinants and Consequences. In: K. C. Zachariah & S. IrudayaRajan (Eds.), New Delhi, Sage publications.

Teng, Y. P. (2007). Living arrangement among the elderly in South East Asia, Paper Presented in the Seminar on the Social, Health and Economic Consequences of Population Aging in the Context of Changing Families, July 25–27, Bangkok, organised jointly by the United Nations Economic and Social Commission for Asia and the Pacific, United Nations Population Fund and United Nations Department of Economic and Social Affairs.

The Hindu. (2011). Kerala has reasons to smile, Meera Prasad, October 1, 2011 edition.

United Nations. (2001). Living arrangements of older persons, Population Bulletin of the United Nations Special Issue No. 42/43.

United Nations. (2005). Living arrangements of older persons around the world, Department of Economic and Social Affairs, Population Division, New York.

Wilmoth, J. (1998). Living arrangement transitions among america's older adults, The Gerontologist 38.

Chapter 7
Socio-Economic Condition and Social Support Among the Ageing Tiwas of Assam

Chandana Sarmah and Barnali Das

Abstract Ageing is a universal phenomenon. But how an individual will age is population and environment specific. The role and status of an ageing person in the society is determined to a large extent by the culture specific to the population. The Tiwas of Assam are a population of Mongoloid origin. Their social organization is patrilineal, and their economy is based on agriculture. The present study was conducted among two villages of Morigaon district of Assam to understand their socio-economic condition and social support system in relation to the ageing Tiwas. 268 male and female Tiwas, above the age group of 50 years were selected. Data were collected with the help of a structured schedule, interview and observation method. The Tiwas are educationally and economically backward. The family is the main source of support for the ageing Tiwas. Joint family is still the dominant type of family among them. Not a single ageing Tiwa person was found to be living alone. Traditional system of keeping resident son-in-law in case a couple does not have a son ensures both social support and residence with at least one adult child among the elderly.

Keywords Ageing · Tiwas · Socio-economic condition · Social support

Introduction

Ageing is a universal phenomenon. The process of ageing involves the interaction of four developmental forces. These are the biological forces, psychological forces, socio-cultural and life cycle forces. Each of us is a product of a unique combination of these forces. These forces create people's developmental experience (Cavanaugh

C. Sarmah (✉) · B. Das
Department of Anthropology, Gauhati University, Guwahati 781014, Assam, India
e-mail: chandanasarmah23@gmail.com

B. Das
e-mail: barnalidas93@gmail.com

B. Das
Department of Anthropology, North Gauhati College, Guwahati, India

S. Irudaya Rajan and G. Balagopal (eds.), *Elderly Care in India*,
DOI 10.1007/978-981-10-3439-8_7

and Fields 2006). Moreover, it is the socio-cultural factors or forces which provide the overall context as to how an individual develops. The culture will shape the core beliefs in people and also provide status, social setting and living condition for people of all ages.

As we all know ageing brings along with it a decline in biological capacities. This decline in biological capacities brings about limitations in a person's ability to cope with his or her environment. As a result, the individual has to depend more on social resources to carry on his day-to-day life. The extent to which these social resources can be utilized is determined by the culture of the population. Ageing, though an individual phenomenon is very much population and environment specific.

This paper is an attempt to understand the socio-cultural factors of ageing among the Tiwas on the basis of their socio-economic condition and prevalent social support system. There are a number of tribes and communities living in the plains and hills of Assam. Each population has their own unique culture, food habits and adaptive process. The Tiwas are a tribe of Assam who are believed to have migrated from Tibet. After reaching the plains, they have moved along the south bank of the course of the river Brahmaputra (Bordoloi et al. 1987) and settled in different places of the state. Their main concentration is found in the undivided Nagaon district in central Assam. There are Tiwa inhabitants even in other districts of the state such as in Lakhimpur, Dhemaji, Jorhat and Kamrup. They are of mongoloid origin and live in both the hills and plains of Assam. The plain Tiwas are patrilineal, and those living in the hills are matrilineal. This may have occurred as a result of acculturation of the Tiwas with the plain dwelling caste population, who are predominantly patrilineal.

Material and Methodology

The present study has been conducted in two Tiwa villages of Morigaon district of Assam, namely Sidhabari and Buragaon. The field work for the study on ageing among the tribe was conducted during 2013–2014. The demographic data of the villages were collected by a demographic survey schedule, and a structured schedule was prepared for collecting data relevant to the other aspects of the study. The rationale for selecting the villages is that both the villages still practice the traditional way of life. Though villages are situated close to the district headquarters, the impact of urbanization can be said to be insignificant. Unlike other rural areas of the state, migration of the younger generation is negligible. The Tiwas in the two villages have more or less retained their traditional culture and way of life.

Sample

The two villages have a total population of 2039 individuals of which 1046 are males and 993 are females. There are a total of 381 households in the two villages. For the purpose of the study, people over the age of 50 years have been taken into

Table 7.1 Distribution of the ageing Tiwas by age and gender

Age group (years)	Male (%)	Female (%)	Total (%)
50–59	70 (52.24)	73 (54.48)	143 (53.36)
60–69	33 (24.63)	38 (28.36)	71 (26.49)
≥70	31 (23.13)	23 (17.16)	54 (20.15)
Total	134	134	268

Source Field survey

Table 7.2 Distribution of the total Tiwa households by their family size

Number of members	Number of households	Percentage
0–3	59	15.49
4–7	274	71.92
8–10	32	8.40
>11	16	4.20
Total	381	100

Source Field survey

consideration, which includes 134 males and 134 females, constituting 13% of the total population. The age–sex distribution of the 50 years and above group individuals is presented in Table 7.1.

The information on age distribution shows that 53.36% of individuals belong to the 50–59 years age category, 26.49% to the 60–69 years and the remaining 20.15% are 70 years and older. With increase in age, the proportion of both men and women gradually come down. In the 50–59 years age category, the percentage of women is higher than that of men. But in the highest age group, the percentage of women is relatively lower than men.

Socio-Economic Condition of the Ageing Tiwas

The family is the primary social unit in any society. Among the Tiwas, it is no different. The family size is not very large, with medium sized families (ranging between 4 and 7) being the most common (Table 7.2).

Marital Status

It is one of the important demographic characteristic in any population and is especially significant in the study of ageing. Having another person to share his/her life with and to care for can enhance the quality of life for elderly people. The marital status has been categorized into four types. First are those who have remained unmarried, second are those who currently have a surviving spouse. The third category includes those who have married but lost their spouse, and the fourth are those who had married but are no longer living with their spouse. A small

Table 7.3 Distribution of the ageing Tiwas by their socio-economic variables according to age group

Age group (years)	Gender	Marital status				
		Never Married	Married	Widow/widower	Divorce/separated	Total
50–59	M	2 (2.86)	60 (85.71)	7 (10.0)	1 (1.43)	70
	F	5 (6.85)	41 (56.16)	25 (34.25)	2 (2.74)	73
60–69	M	0	29 (87.88)	4 (12.12)	0	33
	F	2 (5.26)	18 (47.37)	18 (47.37)	0	38
70+	M	0	22 (70.97)	9 (29.03)	0	31
	F	2 (6.45)	6 (19.35)	15 (65.22)	0	23
Total	M	2 (1.49)	111 (82.84)	20 (14.93)	1 (0.75)	134
	F	9 (6.72)	65 (48.51)	58 (43.28)	2 (1.49)	134

Age group (years)	Gender	Educational level					
		Illiterate	Primary School	Middle School	Completed school	Studied in college	Total
50–59	M	29 (41.43)	7 (10.0)	15 (21.43)	14 (20.0)	5 (7.14)	70
	F	51 (69.86)	10 (13.70)	5 (6.85)	7 (9.59)	0	73
60–69	M	18 (54.55)	7 (21.21)	4 (12.12)	3 (9.09)	1 (3.03)	33
	F	28 (73.68)	4 (10.53)	4 (10.53)	2 (5.26)	0	38
70+	M	13 (41.94)	8 (25.81)	6 (19.35)	3 (9.68)	1 (3.22)	31
	F	20 (86.96)	2 (8.70)	0	1 (4.35)	0	23
Total	M	60 (44.78)	22 (16.42)	25 (18.66)	20 (14.93)	7 (5.22)	134
	F	99 (73.88)	16 (11.94)	9 (6.72)	10 (7.46)	0	134

Age group (years)	Gender	Occupation of the male Tiwas				
		Agriculture	Petty trade	Wage earner	Service	Total
50–59	M	24 (34.29)	8 (11.43)	24 (34.29)	14 (20.0)	70
60–69	M	19 (57.58)	3 (9.68)	10 (30.30)	1 (3.03)	33
70+	M	20 (64.52)	3 (9.68)	3 (9.68)	5 (16.13)	31
Total	M	63 (47.0)	14 (10.45)	37 (27.61)	20 (14.92)	134

Age group (years)	Gender	Income group			
		Low	Middle	High	Total
50–59	M	19 (27.14)	47 (67.14)	4 (5.71)	70
	F	25 (34.24)	44 (60.27)	4 (5.49)	73
60–69	M	12 (36.36)	19 (57.58)	2 (6.06)	33
	F	14 (36.84)	21 (55.26)	3 (7.89)	38
≥70	M	10 (32.26)	17 (54.84)	4 (12.90)	31
	F	9 (39.13)	9 (39.13)	5 (21.74)	23
Total	M	41 (30.60)	83 (61.94)	10 (7.46)	134
	F	48 (35.82)	74 (55.22)	12 (8.96)	134

Note Numbers in parenthesis are percentages
Source Field survey

proportion (4%) of the ageing Tiwas have remained unmarried (Table 7.3). More women than men are found to have remained unmarried among them. Most of the ageing Tiwas are currently married. But if we look into the gender difference, the number of males who have a surviving spouse is significantly higher than females. The number of individuals, both male and female, who no longer have a surviving spouse increases with increase in age, but the proportion of such women is relatively higher than men. In the oldest age group, the number of widows is found to be the highest. This may be an indirect indication of women surviving to higher ages than men.

Education

The educational level of the ageing Tiwas is low. Only 11% of the Tiwas in the 50 years and above age group have completed school education (Table 7.3). Illiteracy is high (59%) and is much higher among women than men in all the age groups. Very few individuals have studied beyond school. The educational level of the children has also been looked into to understand change in the educational level. Illiteracy, though significantly lower, is still found to be prevalent among the succeeding generation. There is high level of school dropouts. Nearly half (52%) of the children have completed school, and among them only 4% have completed graduation (Table 7.4). The proportion of girls completing school is higher than boys. Though there is an increase in literacy level, improvement in terms of educational level between the two generations is not seen. More number of children has completed school relative to the parental generation, but when we look at occupation, we find that it could not have translated into better occupational opportunities.

Table 7.4 Distribution of the children of the ageing Tiwas by their education and occupation

Gender	*Educational level of children*						
	Illiterate	**Primary School**	**Middle School**	**Completed school**	**H.S**	**Graduate**	**Total**
M	13 (6.16)	26 (12.32)	62 (29.38)	77 (36.49)	25 (11.85)	8 (3.79)	211
F	2 (2.63)	12 (15.79)	14 (18.420)	33 (43.42)	12 (15.79)	3 (3.95)	76
	Occupation of the male children						
	Students	**Agriculture**	**Wage earner**	**Petty trade**	**Service**	**Professional**	**Total**
M	52 (24.64)	55 (26.07)	50 (23.70)	38 (18.0)	14 (6.64)	2 (0.95)	211

Note Numbers in parenthesis are percentages
Source Field survey

Occupation

The occupation of a population is a direct indicator of their economic condition. The occupation of the Tiwa population has been looked into from three aspects. One is the past occupation of the ageing population and secondly, to what extent were the ageing individuals involved in economically gainful activities. The third aspect is the occupation of their children. Agriculture is the primary occupation for both the generations. Those who did not have agricultural land of their own, worked as agricultural labourers, were involved in semi-skilled jobs such as masonry, carpentry and electrical works or were rickshaw pullers. Some of them were involved in petty business such as running grocery shops, animal husbandry or selling country liquor. A very small number of males were in service such as teaching, fourth grade government jobs, etc. Women are predominantly housewives and do not have an independent source of income. If we look into the occupational pattern from the point of view of age, then we see that majority of the males of the oldest age group were involved in agriculture. As we move towards the younger ages, the proportion of people being involved in agriculture gradually declines. In the 50–59 years age category, the percentage of males involved in agriculture and wage earning is equal (Table 7.5). In this age group, the proportion of people involved in service is also higher than the other ages. Thus, we can say that with time, there is an indication of people looking for an alternative or additional income source.

The people, who were farmers, continue to work in fields as long as it was physically possible for them to work. It is only from the eighth decade that they stop working in the fields. Individuals who had not gone to the field for work in the last six months have been included in this category. The others go to the field at least thrice a week, but worked according to one's capacity. Those who were involved in petty trade continued with their business with some assistance from family members. Wage earning requires a more rigorous routine than working in one's own field. Therefore, those who were wage earners had to give up working earlier. They were found to be no longer able to earn beyond the seventh decade. Among the individuals working in the formal sector, the age of retirement is 60, after which they are no longer in service.

To get a clear picture about their economic condition, the occupation of the children has also been looked into. No significant change is observed in the occupation of the succeeding generation. Agriculture continues to be the main occupation followed by wage earning. In wage earning, the type of work is found to be different among the younger generation. The transport and communication around the village has improved. The younger generation has a preference to work in this sector. They were working as drivers, mechanics and handyman in the transport sector. Very few were able to find employment in the service sector mainly because of their educational status. Only a couple of children have become professionally qualified.

Table 7.5 Distribution of the ageing population as per their ability to remain working

Age group (years)	Agriculture		Petty trade		Wage earning		Service	
	Continuing to work	No longer working	Continuing to work	No longer working	Continuing to work	No longer working	Continuing to work	No longer working
50–59	24 (100)	–	8 (100)	–	24 (100)	–	14 (100)	–
60–69	19 (100)	–	3 (100)	–	8 (80)	2 (20)	–	1 (100)
≥70	4 (20)	16 (80)	3 (100)	–	–	3 (100)	–	5 (100)

Note Numbers in parenthesis are percentages
Source Field survey

As has been mentioned earlier, the number of school dropouts was higher, especially among the boys. After dropping out, they seek employment in and around the village and continue to live with their families. As a result, there is not much migration of the younger generation out of the village. Another feature that has been observed is that girls though better educated than boys do not seek employment. They remain within the household and thereby do not contribute income to the family.

Income

The income or the economic condition of the ageing persons has been assessed on the basis of the annual family income. The categorization of income has been done on the basis of the Planning Commission poverty estimates (based on consumption expenditure data) for 2011–2012 (released in July 2013), calculated as per the Tendulkar methodology. The Tendulkar poverty line for 2011–2012 was consumption expenditure of Rs. 828 per capita/month in rural Assam. As the average family size of the sample was found to be 5, the average consumption expenditure which was taken as the income has been calculated by multiplying the BPL consumption expenditure by 5 for 12 months.

Therefore, families having an annual income below 50,000 were taken to be below the poverty line. For the study, three income categories have been classified. The first is those who fall below the poverty line. Secondly, those who have an annual income between 50,000 and 3 lakhs rupees has been categorized as the middle income group, and those who have an annual income above 3 lakhs is taken as the high-income group. In addition to the monetary income, the annual produce from cultivation has been converted to rupees.

Nearly one-third (33.21%) of the ageing Tiwas fall below the poverty line, 58.58% have an annual income higher than 50 thousand but less than 3 lakhs, and only 8% have an income higher than 3 lakhs (Table 7.4). The number of ageing women whose families fall below the poverty line is higher than men. This is especially found in the youngest and the oldest age group. In the middle income group, the number of ageing males is higher than females in all the age groups. The males in their sixth and the seventh decade are mostly found to be still involved in economically gainful activities. Their income is supplemented by income of their children. Women, on the other hand, do not have any income of their own but are dependent either on their spouse or on their children. Most of the ageing women as has already been mentioned do not have a surviving spouse. This may be a reason for their families facing economic hardships. Among those who fall in the high-income category, the proportion is found to be highest in the oldest age category. In this situation, we can conclude that their economic condition has improved mainly from the income of the children.

Social Support Among the Tiwas

Individuals do not just grow old, but they grow old in a social and cultural context. Social institutions organize social relationships and determine role and functions. This in turn formulates the social life of the members. Krishnaswamy and Shanthi (2010) says that an adequate system of social support is an integral aspect of optimal health in old age in addition to freedom from illness, optimal functional status and continuing personal development. This is very rightly said, as it is known that the psychological aspects of well-being and quality of life are determined largely by the type of relationship that an elderly person has. The biological, social and psychological aspects are very much interrelated to the life cycle forces in the process of adult development and ageing. It is the culture and society of a particular community that will determine the role and status of the elderly people. In rural areas, the provision of services is practically impossible for a country like India. But what is most interesting is the traditional culture, wherein irrespective of the tribes or the caste population, there is a natural support network. This support network operates from the social structure, which ensures a support for both the aged and disabled. In places where urbanisation has not taken place or the impact of globalization has not percolated, the natural support networks are still in operation.

The Tiwas are one such community where the natural support system are by and large still in operation. As has been mentioned in the discussion on the socio-economic condition of the ageing Tiwas, we find that the educational level and occupation has not changed much between the ageing generation and their children. Migration for work or education among the younger generation is almost negligible. As such the traditional way of life continues.

According to Fry (2013), to provide security for either younger or older members of society, there must be a combination of both resources and rewards. Some families and some older persons will have more wealth than others. Wealth not only has to be managed but attention has to be given to who will inherit it once a person dies or become disabled. Protection is also provided to older people in explicit rules of co-residence and inheritance of family wealth. Family units provide security, and they are also the units of production. Older people are rarely excluded from division of labour, because they are needed for their experience, knowledge, work connections and sometimes wealth. In the present study, social support has been examined from three aspects. One is the living arrangement of the ageing individuals. Secondly, the role of the ageing persons in the social life of the community has been looked into and finally their role in the family.

Living Arrangement of the Ageing Tiwas

Among the Tiwas of Assam, joint or extended family is the most prevalent type of family. Ageing individuals live with their married or unmarried children. There is also prevalence of the tradition of having a resident son-in-law. A section of Tiwas live in the hills and are known as the hill Tiwas. The plain Tiwas have migrated from the hills and settled in the plains. In the plains, they got acculturated with the plains peasants and adopted the Vaishnav religion. The hill Tiwas are matrilineal, and the plain Tiwas have adopted a patrilineal social organization. The custom of having a resident son-in-law may be a practice, which has continued from their matrilineal origin. Among the plain Tiwas, one of the daughters along with her husband continues to live in the maternal home after marriage. This happens when the family has a large amount of land or do not have a son. This ensures care and support for the ageing parents. In case of transmission of property, the resident son-in-law is given an equal share as the sons. If there is no son, then the property is inherited by the son-in-law. However, with a change in the occupational pattern, this tradition is slowly disappearing. Sometimes the son-in-law may be employed in another place, which does not permit him to stay with his in-laws. In course of the study, only 2.24% of the ageing males and females were living with their resident son-in-law. Married sons along with their families continue to live in their parental home as a mark of filial piety. If a couple has more than one son, then after all the sons get married, the cultivable land is divided equally among the sons. They can now set up independent homes and the son who looks after the parents inherits the homestead land. In case there is an unmarried daughter in the family, then an equal share is given to her. She will continue to live with her parents or any of the brothers. Whoever will take care of her will inherit the share of her land. The unmarried males and females found in the study were living with their brother's family. Individuals who did not have children or were not living with their spouse were taken care of by their nephews. A very small percentage (2.61%) of the ageing individuals were living with only their spouse. Another very noteworthy feature in the living arrangement was that none of the ageing Tiwas were found to be living alone. Classic modernization theory (Cowgill and Holmes 1972) view living arrangement as one determinant of well-being. It assumes that elders who live with adult children or grandchildren are better provided for and hence enjoy higher status. This is found to be true to a large extent among the Tiwas. The elderly especially women do not have money or any purchasing power in their hands but appear to be quite content. The son or other family members provide for whatever they need. Their expectation level is not very high, and their needs are also small (Table 7.6).

Elderly who are above 60 years of age are found to require assistance in their activities of daily living. The level of dependence is higher among women. The assistance required is given by the family members mostly the daughter-in-law. In case there is an unmarried daughter, she takes care of the needs of the parents. But after her marriage, the daughter-in-law takes up the responsibility. This is taken as

Table 7.6 Distribution of the ageing Tiwas by their living arrangement

Age Group (years)	Gender	With only spouse	With Spouse, unmarried children	Widowed with unmarried children	With spouse in extended/joint family	Widowed in extended/ joint family	With resident son-in-law	Brother's family
50–59	M	4 (5.71)	30 (42.86)	5 (7.14)	18 (25.71)	9 (12.86)	2 (2.86)	2 (2.86)
	F	1 (1.37)	18 (24.66)	5 (6.85)	22 (30.14)	21 (28.77)	1 (1.37)	5 (6.85)
60–69	M	1 (3.03)	10 (30.3)	–	18 (54.55)	54 (12.12)	–	–
	F	1 (2.63)	2 (5.26)	–	15 (39.47)	18 (47.37)	–	2 (5.26)
≥70	M	–	1 (3.23)	–	20 (4.52)	77 (22.58)	1 (3.23)	2 (6.45)
	F	–	–	–	7 (30.43)	12 (52.17)	2 (8.70)	2 (8.70)
Total	M	5 (3.73)	41 (30.60)	5 (3.73)	56 (41.79)	20 (14.93)	3 (2.24)	4 (2.96)
	F	2 (1.49)	20 (14.93)	5 (3.73)	44 (32.84)	51 (38.06)	3 (2.24)	9 (6.72)

Note Numbers in parenthesis are percentages
Source Field survey

an expected role to be fulfilled by the daughter-in-law. As and when necessary, she is assisted by the sons and even the grandchildren. The family structure enhances in overcoming the effects of disability in elderly. This in turn ensures feelings of well-being among the elderly.

Role of Elderly in Social Life

The objective of examining the role of the ageing individuals in society is to understand whether they are socially active or disengaged. Remaining active or having a role in society is necessary for well-being in later years. The social life of a tribal community mainly revolves around their religious beliefs and practices and life cycle rites and rituals.

The Tiwa community in course of their migration from hills to the plains have not only adopted the peasant way of life, but also accepted Vaishnavism. In the two study villages, followers of both Vaishnavism and the traditional Tiwa religion were living together. The traditional Tiwa religion is based on the belief of idols and shrines. Followers of both the religions have a few community functions, which they perform annually. In addition to this, there is an annual fair which is not related to religion but a celebration of the Tiwa culture. Both the factions take part equally. The fair is organized mainly by the youth where Tiwa dance, music, songs, along with other fanfare are organized. The elderly visit the fair, but there is no significant role for them. But in the other rituals of both the religious groups, the elderly occupy important positions. Those who follow Vaishnavism organize their religious functions in the "*NaamGhar*" or a community prayer hall. From amongst the followers, few persons are appointed for conducting different duties. The elderly persons generally supervise the work and advise the workers. In the Assamese calendar month of *Bhado,* which falls in the second half of the English month of August and the first half of September, prayers are held everyday. Each clan takes the responsibility of organizing the prayers for one week and follows the routine in a cyclic manner. The women, especially the elderly women, play an active role in these prayers. Preparation of the offerings, conducting the prayers and distribution of the offerings are carried out by the senior and experienced women. Certain code of conduct has to be followed in performing these duties, and they are strictly followed. During other months, the community prayer halls are looked after and maintained usually by an elderly woman. She may be assisted by a family member, mostly the daughter-in-law. The elderly lady is also responsible for lighting the lamps every morning and evening.

Among the people who follow the traditional Tiwa religion, the role of the elders is very significant. For the rites of birth, marriage and death, the elders have a specific role. In case of birth and attainment of puberty, there is no specific ritual. Births generally take place in the homes. A midwife is called for and elderly women are present to give advice. In the traditional marriage system, after the ceremony, the bride and groom are ritually blessed by the village elders. The elders sit on one side, and blessing is given by chanting of hymns in the Tiwa language.

Similarly, in case of death, after the cremation and official mourning period, when family and clan members have to follow restrictions and taboos, a few elderly persons are invited for ritualistic food offerings by the family of the dead. The actual ritual of mourning may be delayed by 5 or 6 years depending on the economic condition of the family.

Each village has a common cremation ground, and each clan has a specific place for cremation. These places are in the form of mounds or small hillocks. Clan members are cremated in their respective mounds after the observation of certain omens. The mound for a particular clan has to be identified by a senior clan member.

The followers of the traditional Tiwa religion perform a few other rituals. The Tiwas believe in the existence of '*thans*' or sacred groves, where rituals are performed. These rituals are related to either ancestral worship or the general welfare of the village. Each village may have its own sacred '*than*' or a number of villages may have one common place. One such specific place is called an '*aai than*', where an annual ritual is conducted only by elderly women. Strict rules of prohibitions and taboos are meticulously followed while performing the ritual. It is performed for the general welfare of the village. Another ritual known as '*Jhongkhong puja*' is also performed by the senior members of the community. This is performed by the male members.

Role of the Elderly in the Family

Fry (2013) has very rightly pointed out that in rural and traditional way of life, the elderly were not excluded from the division of labour but very much included in the system for their knowledge, wisdom and experience. Men transfer some of the authority to his sons after they mature. In cultivation they follow the age old system and as such rely on experience. But this is not found to be the case with new types of occupation. Here the classical theory of modernization (Cowgill and Holmes 1972), which says that a shift from the traditional occupation brings down the status of the elderly, may prove true when the change is on a bigger scale. When a Tiwa male is no longer actively involved in agriculture, he uses his time to work in the kitchen garden or make articles of bamboo. He is hardly seen to be sitting idle. During the day, he carries a knife or some implement in his hands for work. Most of the articles of day-to-day use among the Tiwas are made from bamboo, which is available in plenty in the village. The Tiwas are very fond of fishing and every year during winter, they have a communal fishing festival where males and females of all ages participate. The fishing implements are made from bamboo. Moreover, the musical instruments used in their dance and songs are all made of bamboo. It is only a few handful of elderly males who are experts in making them. Between the two study villages, there is only one man who can make such musical instruments.

Among the women, after the son gets married, the mother gradually hands over the responsibility of the kitchen to the daughter-in-law. The mother-in-law's role becomes more supervisory and giving advice. After the birth of grandchildren, the grandmother plays an important role in childcare activities. Grandmothers generally keep toddlers occupied through play, allowing the daughter-in-law to carry on the household chores. All the other children of the village also refer to elderly persons as they do their own grandparents, and the seniors also treat them likewise. Reprimanding a child for some mischief is not taken as an offence by the family. The grandparents treat grandchildren with affection and are in many cases, their sought-after companions.

Elderly males have a say in decision making but the females are not found to play a very decisive role. But what is noteworthy is that expectation of the elderly females in decision making is not much. The general opinion is that their sons are now mature enough to make a decision, and he will do so according to his capacity and ability. Even when a child drops out of school, the parents do not force him/her to attend school.

Conclusion

The Tiwas of the study area, despite living close to the district headquarters, follow a traditional rural way of life. They practice agriculture as their primary means of living, though a shift towards other means of subsistence is also observed. The educational level of the younger generation being low prevents them from migrating out of the village for better occupational opportunities. This has, in a way, worked to maintain the family structure and the support system within the family. The social structure ensures a high status for the elderly. Taking care of one's ageing parents and relatives are considered the responsibility of the children, and they fulfil this without any question. The social structure ensures that the effects of impairment and disability among the aged or any other section of the society are taken care of by the family. Their simple way of life and thinking ensures a satisfactory living for all age groups. However, modernization and subsequent changes may completely alter the social situation.

References

Bordoloi, B. N., Thakur, G. C. S., & Saikia, M. C. (1987). *Tribes of Assam, Part I.* Assam: Tribal Research Institute.

Cavanaugh, J. C., & Fields, F. B. (2006). *Adult development and ageing* (5th ed.). USA: Wadsworth Thompson Learning Inc.

Cowgill, D. O., & Holmes, L. D. (1972). *Ageing and modernization.* New York: Appleton-Century-Crafts.

Fry, C. L. (2013). Social anthropology and ageing. In D. Dannefer & C. Phillipson (Eds.), *The sage handbook of social gerontology*, Sage Publication Ltd. London.

Krishnaswamy, B., & Shanthi, G. S. (2010). Health promotion and protection. In S. I. Rajan & C. S. Johnson (Eds.), *Ageing and health*. Jaipur: Rawat Publication.

Chapter 8
Care and Support During Twilight Years: Perception of Elderly from Rural India

Dhananjay W. Bansod

Abstract In today's context of modernization and its associated developments, it is necessary to know not only the perceptions of elderly about the care and support provided by their children but also those of the younger generation with a view to understanding the current situation particularly in rural areas. Present paper focuses on care and support that elderly gets in family and their perceptions towards the younger generation in rural Maharashtra. Issues like respect and care, how the elderly and their subsequent generation perceive ageing, elderly's views and opinions about young generation have been explored in this study. Further, the impact of various socio-economic characteristics on the perceptions of elderly towards younger generation is discussed in detail. Data for this study were collected from rural Maharashtra using semi-structured interview schedule. A total of 600 males and females were interviewed using systematic sampling technique and use of multivariate analysis. Result reveals that one-third of elderly in rural area were not getting proper care and support from their children/family and many of them perceived ageing as a problem. The study highlights the need for company of either married or unmarried children to make elderly feel more secure.

Keywords Elderly · Perception · Younger generation · Maharashtra

Introduction

It is commonly known that in old age, individuals depend on their children, as they often have no other alternative. The effects of modernization, industrialization and urbanization on the society are evident, in part, in the weakening of traditional bonds of the joint family system. This breakdown of the joint family system is more common in urban areas, as it is evolving increasingly into a nuclear family system.

D.W. Bansod (✉)
Department of Public Health and Mortality Studies, International Institute for Population Studies, Deonar, Mumbai 400088, Maharashtra, India
e-mail: dbansod@gmail.com; dhananjay@iips.net

S. Irudaya Rajan and G. Balagopal (eds.), *Elderly Care in India*,
DOI 10.1007/978-981-10-3439-8_8

The elderly happen to be the main sufferers of the changing social values and family system (Gaur and Kaur 2004). In today's context of modernization and its associated developments, it is necessary to know not only the perceptions of the elderly about the care and support provided by their children but also those of the younger generation with a view to understanding the current situation particularly in rural areas.

The spread of consumerism and self-centric attitude of the younger generation have driven them to drift away from their villages in search of comfort and better opportunities in urban areas. In such a situation, the poor and helpless aged parents are left behind alone in the villages, where they feel socially isolated and economically insecure, compelled to lead a life of uncertainty and difficulty (Behara and Mohanty 2005).

A study by Asharaf (2000) in Kerala found that there were elderly people who thought that today's youth were disrespectful to elderly, considered them a burden, and a hindrance to planning outings, and that they do not give importance to their feelings or they physically abuse them, besides feeling helpless about their being aged. A majority of the elderly without financial/religious engagements perceived ageing as a problem. Similarly, the likelihood of perceiving ageing as a problem is found to be higher among those who do not have frequent contacts with relatives and friends, being forced to handle depression all by themselves, or those who do not enjoy the status as the head of the household and those who are not consulted in the decision-making process of the family (Asharaf 2000).

In this context, the present study aimed to examine the issue of old age from the perspective of the elderly and the young generation (their children). The study also tries to understand the perception of the elderly in terms of respect and care they get from their children.

Objectives

1. To understand the issue of the elderly in the context of respect and care.
2. To assess the perceptions on ageing from the point of view of elderly and the younger generation (their children).
3. To study the perceptions of elderly people towards the younger generation.

Study Area

In Maharashtra, the elderly population, which was 5% in 1961, increased to 9% in 2001. This is an evidence of the increase in the proportion of elderly people in the total population of Maharashtra. The increase also indicates an overall improvement in the field of public health care in India. Rural–urban distribution shows that there are more elderly people living in rural areas as compared to urban areas. According to the 2001 census, 10% of the population in rural areas were elderly (60+), while the elderly constitute 6.8% of the population living in urban areas. This variation is a topic of interest for studying the issue of ageing in rural Maharashtra. According to 2001 census, 11% of elderly are found in rural areas of Amravati district, which was very high when compared to the national average (i.e. 7.4%), and also a majority (71%) of the elderly population were living in rural areas (Census 2001).

Design of the Study

Amravati district has been selected purposively, and further, a three-stage sampling design was adopted with the selection of blocks in the first stage, villages in the second stage and households in the third stage. During the survey, we collected information from 600 respondents in 15 villages of Amravati district. One block was selected at random out of 14 blocks in Amravati district in the first stage, and all the villages in the selected block were then classified into three categories on the basis of population size, i.e. less than 500, 500–1000 and more than 1000. Fifteen villages were selected in the final stage from the three categories proportionate to the total number of villages in the particular category. A sampling frame of the aged 60 and above was then prepared for the selected villages by house listing. This was followed by the selection of elderly respondents using systematic random sampling method. The survey was conducted between January and April 2006.

The data were collected using pre-tested, semi-structured interview schedule. Detailed information was collected regarding basic socio-economic variables: situation of the elderly in the context of respect and care, how the elderly and the younger generation (their children) perceive ageing and the elderly's views and perceptions about the young generation. Further, the impact of various socio-economic characteristics on the perceptions of elderly people on younger generation is also discussed in detail. Chi-square test is used for establishing associations. Multivariate analysis is also used to examine the impact of predictor variables on the perceptions of the elderly about the younger generation.

Results

Respect and Care in Old Age

Questions were asked of elderly respondents in terms of respect and care they provided to their parents and examined whether they felt they were getting the same respect and care from their children at present and 37% of the elderly reported that they were not getting as much respect and care from their children that they had provided to their own parents (Fig. 8.1), whereas 59% of the elderly reported that they do get the same kind of respect and care from their children as had been given

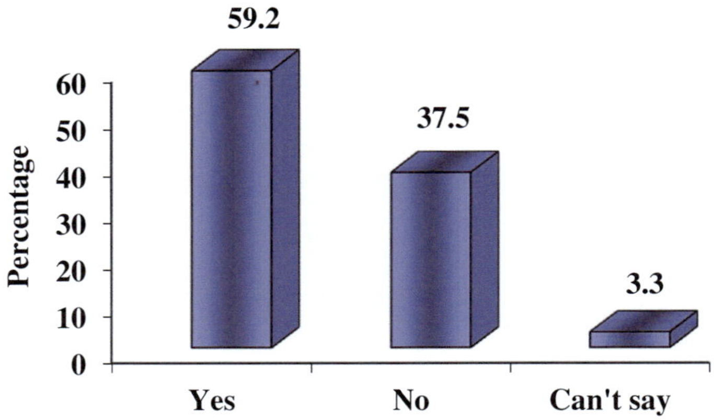

Fig. 8.1 Elderly receiving respect and care from children

to their own parents; however, 3% of the elderly are found to be silent about the care provided by their children.

A possible explanation to this kind of silence is that they were not getting proper care and support or that they could not say anything to anyone because if they complain, they may be denied even the present minimum care and support being given. Further, an attempt was made to look at the background effect on the respondents' opinion regarding their children.

Table 8.1 displays the distribution by background characteristics of elderly people's opinions about the care and support given by their children as compared to their own parents. By looking at categories of age groups, it becomes clear that more respondents from the 'oldest of the old', namely those aged 80 and above years (94%), group reported that their children provide as much respect and care as they used to provide to their own parents. However, more elderly respondents from the 'young–old', namely those aged 60–69 years (42%), felt that they were not getting as much respect and care from their children as was provided to their parents. This kind of perception could be taken as an example of the generation gap, societal and value differences between parents and children, which has been on the increase over the last 10–20 years due to modernization and urbanization. The elderly from the oldest of the old group have reported that they get more care and respect from their children because of the fact that their children are relatively older than the children of the elderly from the young–old group. The generation gap between the oldest of the old and young–old is large. A decade or two ago, children were more likely to show respect and care to their parents in old age, treating them as gods, but this situation has changed over time and is reflected in the way the elderly from the young–old age group are not getting proper care and respect from their children, as they had provided for their parents.

Sometimes it is found that after the son's marriage, it becomes very difficult for parents to adjust with their son's new family. Often elderly men adjust more easily

Table 8.1 Percent distribution of elderly according to their children giving as much respect and care as they did for their parents by selected background characteristics

Background characteristics	Children giving as much respect and care as you did for your own parents			Total
	Yes	No	Can't say	
Age group				
60–69	55.3	41.5	3.2	347
70–79	59.8	36.0	4.2	189
80+	94.3	5.7	–	35
Sex				
Male	61.3	35.1	3.6	305
Female	56.8	40.2	3.0	266
Marital status				
Married	63.1	33.4	3.5	344
Widowed/widower	53.3	43.6	3.1	227
Living arrangements				
Living alone	33.8	57.7	8.5	71
Living with spouse	45.0	45.9	9.0	111
Living with children	67.5	31.7	0.8	379
With other relatives	80.0	20.0	–	10
Type of family				
Single	25.5	64.7	9.8	51
Nuclear	49.8	46.7	3.5	259
Joint	75.1	23.0	1.9	261
SLI				
Low	51.0	43.4	5.6	198
Medium	65.2	32.1	2.7	224
High	61.1	37.6	1.3	149
Economic status				
Independent	53.8	41.2	4.9	325
Dependent	66.3	32.5	1.2	246
Total	59.2 (338)	37.5 (214)	3.3 (19)	100.0 (571)

Source Field survey

because they do not have as much direct contact with the daughter-in-law as the female household members who are at home most of the day. In the new family setting, the power goes to the daughter-in-law and the son often neglects his parents because he is busy with his new family (Fig. 8.2). The elderly people living alone reported that they do not receive respect and care from their children as compared to the elderly living with children in joint families.

A large proportion of elderly respondents with a low standard of living and who were economically independent felt they did not get care and respect they deserved from their children as compared to those with a high standard of living and

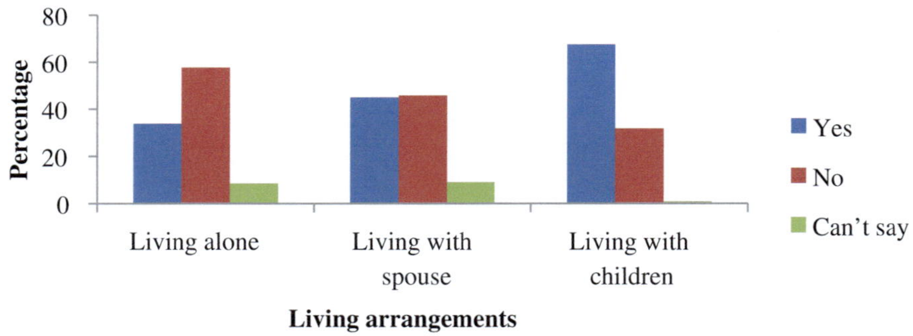

Fig. 8.2 Elderly receiving care and respect from children by living arrangements

dependent status. Often if the elderly live with their children, it means their children take care of them as well as show respect. On the other hand, children of elderly parents with a low standard of living may not be able to afford their parents' care and thus are unable to keep their parents at home with them, forcing the parents to fend for themselves and become financially independent. About 60% of elderly respondents felt that their children are providing them with as much respect and care as they used to give their own parents, while the remaining feel they were not respected and taken care of as much as they should be.

Views of Elderly About Sons

Table 8.2 displays the elderly respondents' views concerning their sons, in terms of differences by selected background characteristics. To a question on whether having a son is beneficial for them in old age, 80% of respondents have reported 'yes', whereas 20% of the elderly have said 'no'. A large proportion of elderly respondents (82%) in the 'young–old' group (60–69) felt having sons is beneficial for the elderly, while 28% of the elderly in the 'oldest–old' age group (80+) and 22% of elderly in the 'old–old' group (70–79 years of age) reported that having sons was not beneficial. As sons were taking care of their own family, they often left their older parents alone. This transition has happened slowly over time and is an example of the changing social and traditional values and norms in the Indian society.

More male and married elderly respondents view that having a son was beneficial as compared to female and widowed/widower elderly respondents. A large number of elderly respondents living alone and independent with a low standard of living index have reported that having a son was not beneficial for the elderly as compared to the elderly living in joint families with a high standard of living and economically dependent on their children. The possible reason for this observation could be that the elderly who were abandoned by their sons have to live alone and

Table 8.2 Perception of elderly about benefits of having sons by selected background characteristics

Background characteristics	Having sons are beneficial for the elderly		Total
	Yes	No	
Age group			
60–69	81.6	18.4	365
70–79	78.5	21.5	195
80+	72.5	27.5	40
Sex			
Male	81.3	18.7	315
Female	78.6	21.4	285
Marital status			
Married	83.5	16.5	352
Widowed/widower	75.0	25.0	248
Type of family			
Single	62.9	37.1	62
Nuclear	79.5	20.5	263
Joint	84.4	15.6	275
SLI			
Low	71.3	28.7	218
Medium	84.3	15.7	229
High	85.8	14.2	155
Economic status			
Independent	78.7	21.3	342
Dependent	81.8	18.2	258
Total	80.0 (480)	20.0 (120)	100.0 (600)

Source Field survey

fend for themselves. As per the traditional culture, in India sons are expected to look after their parents in old age, this change in social values has driven the elderly into feeling bitter about being left alone. Furthermore, if the elderly have a low standard of living, their poor financial condition may force them into believing that 'having sons is not beneficial for the elderly'.

Elderly's Opinion About the Younger Generation

Figure 8.3 presents the perceptions of elderly respondents towards the younger generations in terms of providing care and support to their elderly parents. Of elderly respondents, 28% reported that the younger generation takes care of their aged parents. However, 22% have stated that children do not take care of their aged

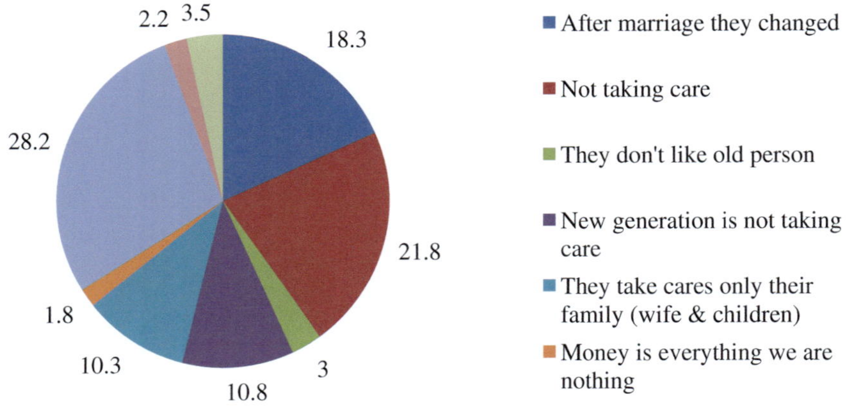

Fig. 8.3 Opinion about the young generation regarding care and support given to elderly parents

parents and 18% of elderly stated that after marriage, behaviour of the younger generation changes. Of the respondents, 10% have stated that the younger generation takes care of their own immediate family and not their aged parents, while the remaining elderly feel that the youngsters do not like old persons/parents and that money is everything for them and old parents mean nothing.

Perceptions of Elderly Respondents Towards Younger Generation

The study questions are formed on gaining a better understanding of the elderly's opinion towards the younger generation concerning particular aspects of their life. Some questions are positive and some negative in nature. To understand the perceptions of elderly towards younger generation, a composite score has been computed based on 12 questions after checking the reliability using the alpha value of 0.8531. The composite score has been computed attaching a higher value to positive indicators and lesser values otherwise (i.e. value 2 for agree, 1 for disagree and 0 cannot say). The scores have been categorized into three groups on the basis of cumulative frequency, neutral (below 25%), positive (25–50%) and negative (above 50%).

Distribution of perception scale indicates that 54% of the elderly respondents had a negative perception towards the younger generation, whereas 28% of the elderly respondents had a positive perception of the younger generation and 18% were neutral (Fig. 8.4).

The perceptions of elderly respondents towards younger generation as compared by their living arrangements are presented in Table 8.3. A larger number of elderly respondents living alone and those living only with their spouses (73 and 62%, respectively) were found to have a negative perception towards younger generation

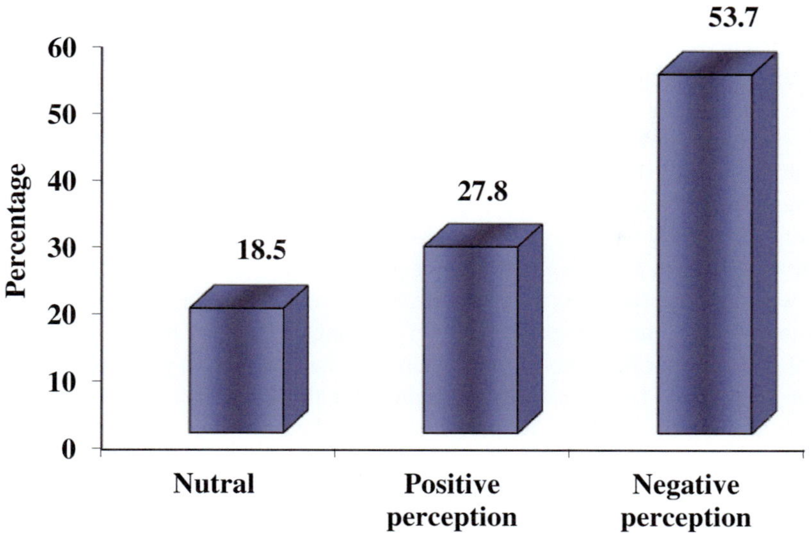

Fig. 8.4 Perception of elderly about young generation

Table 8.3 Perception of elderly towards young generation by living arrangement

Living arrangement	Perception towards young generation			Total
	Neutral	Positive perception	Negative perception	
Living alone	6.5	21.0	72.6	62
Only spouse	13.3	24.2	62.5	120
Spouse + unmarried son	15.6	31.2	53.2	109
Spouse + married son	25.7	27.1	47.1	140
With married sons	26.3	30.8	42.9	133
With other relatives	8.3	33.3	58.3	36
Total	18.5 (111)	27.8 (167)	53.7 (322)	100.0 (600)

Note Chi-square = 30.384 @ $P < 0.001$
Source Field survey

as compared to those living with their children. This negative attitude is found to increase with elderly parents who either stay alone or with their spouses. However, more number of elderly people living with their children and those living with other relatives reported a positive perception towards the younger generation. Of the elderly respondents, 18% had a neutral perception towards younger generations, and among those, 26% were elderly living with spouses and married sons and 16%

were living with spouses and unmarried sons. Having the company of either married or unmarried children makes the elderly feel more secure, and therefore, elderly are more likely to have a positive attitude towards younger generation.

Table 8.4 presents the percentage distribution of elderly respondents according to their perception towards the younger generations by household headship. More than half of the elderly respondents in the study area have a negative perception about the younger generation, while about 30% have a positive perception.

Among the elderly who were not the heads of their households, 44% of them had a negative perception towards the younger generation, whereas among those elderly people who were the heads of their households, about 58% of them had a negative perception towards the younger generation. Among those who happen to be the heads of one-member households, more than 70% had a negative perception of the younger generation. The corresponding figures for those elderly respondents who are heads of households of two members and households of two or more members who had a negative perception were 65 and 51%, respectively.

Table 8.5 shows elderly respondents' differences in perception by background characteristics. Perception among the elderly respondents differs by age; a majority of (70%) the 'oldest of the old' (80+) were found to have a negative perception about the younger generation as compared to the young–old age group; this is a worldwide phenomenon in that as people get older, they tend to need more care and support physically as well as emotionally. If the elderly feel they are not receiving the needed support, they may blame their children for not taking proper care of them and thus develop a negative perception towards the younger generation. More number of uneducated female respondents from scheduled castes/scheduled tribes reported a negative perception towards younger generations as compared to other

Table 8.4 Perception towards younger generation by headship

Household head	Perception towards young generation			Total
	Neutral	Positive perception	Negative perception	
Non-elderly head	25.1	30.6	44.3	183
Elderly head	15.6	26.6	57.8	417
Household head				
Non-elderly head	25.1	30.6	44.3	183
Elderly head in one-member HH	6.5	21.0	72.6	62
Elderly head in two-member HH	11.8	23.6	64.5	110
Elderly head in more than two-member HH	19.6	29.4	51.0	245
Total	18.5 (111)	27.8 (167)	53.7 (322)	100.0 (600)

Note Chi-square = 24.391 @ $P < 0.001$
Source Field survey

groups. Elderly male respondents with education up to primary level from the general category had a positive perception concerning the younger generation. It was also found that more elderly people living alone, with a low standard of living and economically independent, were more likely to feel negative about the younger generation as compared to those living with their families with a high standard of living and economic dependence on others.

Table 8.5 Perception of elderly towards young generation by background characteristics

Background characteristics	Perception towards young generation			Total
	Neutral	Positive perception	Negative perception	
Age groups				
60–69	18.4	28.2	53.4	365
70–79	20.5	28.7	50.8	195
80+	10.0	20.0	70.0	40
Sex				
Male	17.5	30.2	52.4	315
Female	19.6	25.3	55.1	235
Education				
Illiterate	18.1	27.0	55.0	382
Primary	16.0	30.6	53.5	144
Middle school & above	25.7	27.0	47.3	74
Caste				
General	18.0	34.0	48.0	50
SC/ST	16.3	25.2	58.5	313
OBC	21.5	30.0	48.5	237
Marital status				
Married	17.6	28.1	54.3	352
Widowed/widower	19.8	27.4	52.8	248
Type of family				
Single	6.5	21.0	72.6	62
Nuclear	19.4	31.6	49.0	263
Joint	20.4	25.8	53.8	275
SLI				
Low	8.8	27.3	63.9	216
Medium	21.0	24.9	54.1	229
High	28.4	32.9	38.7	155
Economic status				
Independent	17.5	28.4	54.1	342
Dependent	19.8	27.1	53.1	258
Total	18.5 (111)	27.8 (167)	53.7 (322)	100.0 (600)

Source Field survey

Multivariate Analysis

Logistic regression analysis shows the effect of background characteristics on the elderly's perception towards the younger generation in Table 8.6. The dependent variable relates to the perception of the elderly about the younger generation (i.e. 1 = negative perception, 0 = otherwise). After controlling for selected predictor variables, it was found that age, living arrangements and standard of living index significantly affect the elderly's perception towards the younger generation.

It has been observed that the 'oldest of the old' (80 years and above) people are two times more likely to have a negative perception of the younger generation as compared to 'young–old' (60–69 years) people. As age advances, elderly people develop more vulnerability to illness and their health problems increase, and over time, they need more health care. Now it is common that the elderly's children do not have the time to care for them properly. This lack of care makes the elderly develop a

Table 8.6 Logistic regression analysis of negative perception of elderly towards younger generation

Variables	Exp (B)
Age group	
60-69®	
70-79	0.872
80+	2.061**
Sex	
Male®	
Female	1.053
Marital status	
Married®	
Widowed/widower	0.896
Caste	
General®	
SC/ST	1.347
OBC	1.049
Type of family	
Single®	
Nuclear	0.933
Joint	1.642
Living arrangement	
Living alone®	
Living with spouse	0.909
Living with children	0.471**
Living with other relatives	0.670
SLI	
Low®	
Medium	0.740
High	0.422***

(continued)

Table 8.6 (continued)

Variables	Exp (B)
Economic status	
Independent®	
Dependent on others	1.082
Constant	1.868

*** $P < 0.001$, ** $P < 0.005$

negative perception towards the younger generation as compared to the relatively younger elderly. Living with their children has a significant relationship with the elderly's perception towards the younger generation. Elderly people living with children were less likely to have a negative perception towards the younger generation as compared to those living alone, because those who live with their children may receive the expected care and support and hence, their perception towards the younger generation was more likely to be positive. In comparison, those living alone do not get any kind of help from their children, so their perception was more likely to be negative. Further, those with a high standard of living were also significantly less likely to have a negative perception towards the younger generation as compared to those elderly with a low standard of living. This finding comparing the two socio-economic groups could be true because those elderly people with a high standard of living may not expect as much from their children. When those with a higher standard of living receive care and support from their children, they appreciate it, but were not dependent on it, as compared to those with a low standard of living who were dependent on their children's care for day-to-day needs.

The expectations from children by the elderly with a lower socio-economic level are greater as they need more care and support. So when this group does not get the needed care and support, they tend to develop a negative perception towards the younger generation. The remaining predictor variables are not significantly related to the elderly's perception towards the younger generation. A respondent's gender, belonging to scheduled castes/scheduled tribes, living in joint families and being dependent on others are some of the characteristics more likely to influence a person to develop a negative perception towards the younger generation; however, these associations are not statistically significant.

Being Aged Is a Problem

In old age when the earning capacity declines or when the elderly retire from work, the situation could be overall different, in that their assets (money or property) could be transferred to their children partially or wholly. Thus, being dependent on others for everything can make someone, who is familiar with providing for oneself, feel miserable, and further, the elderly may not be able to accept this new development. However, if the elderly have sufficient income to take care of their own day-to-day needs, the situation may be different. Hence, the perceptions of the elderly about ageing as such may differ based on varying life circumstances. Male elderly people during their adult years do not need to ask for anything from others; rather, they are commonly the family provider—sole or otherwise. When the role is reversed in old age, they may find it difficult to get used to a new situation. The parents may feel hurt when their own children neglect their needs and requirements. They generally experience psychological and emotional shocks when faced with such situations.

Table 8.7 shows the perceptions of the elderly who felt that being aged is a problem by selected background characteristics. Of the elderly respondents, 58% perceived being aged as a problem and the remaining did not perceive it as a problem. More than 59% of the elderly in the 'young–old' age group (60–69 years) perceived becoming old as a problem. There is no significant difference in terms of gender regarding the perception about old age. However, more number of widowed/widower elderly (62%) respondents perceived ageing as a problem as compared to married elderly people (55%). It is generally argued that as age advances, women's health needs get neglected due to their so-called lower status and widowhood, which further accentuates their unhappiness (Radha Devi et al.

Table 8.7 Distribution of elderly by their perception (being aged is a problem) and selected background characteristics

Background characteristics	Being aged is a problem		Total
	Yes	No	
Age group			
60–69	58.8	41.2	354
70–79	56.3	43.8	192
80+	54.1	45.9	37
Sex			
Male	58.3	41.7	307
Female	56.9	43.1	276
Marital status			
Married	54.8	45.2	345
Widowed/widower	61.8	38.2	238
Type of family			
Single	69.5	30.5	59
Nuclear	63.7	36.3	256
Joint	49.3	50.7	268
SLI			
Low	60.1	39.9	208
Medium	55.5	44.5	227
High	57.4	42.6	148
Economic status			
Independent	59.3	40.7	334
Dependent	55.4	44.6	249
Total	57.6 (336)	42.2 (247)	100.0 (583)

Source Field survey

1999). A large proportion of elderly respondents living alone (70%) perceived ageing as a problem as compared to those who live in joint families. Those who lived alone did not have anyone to look after them, as they were left behind by their children; further, whenever they need help, even during an emergency situation, they had to seek help from non-family members and this made them feel lonely and frustrated emotionally, which made them perceive ageing as a problem.

More number of elderly people with a low standard of living and economic independence perceived ageing as a problem compared to those with a high standard of living and were dependent on their children. Further, those who were dependent on their children received some emotional and financial support whenever needed unlike those living alone with a low standard of living; they have to work to run their households even when their health does not permit. Hence, for those living all alone, ageing truly does appear to be a problem.

A qualitative analysis of the survey data also reveals that most of the elderly respondents perceived ageing as a problem. There are numerous and varying reasons cited by the elderly for their negative attitude towards the problem of ageing. The most common reasons for perceiving ageing as a problem were that they can no longer do the work they used to do once; health problems increase with age; not getting the expected love and respect from children and other family members; and having to be dependent on others for each and every thing. A 65-year-old female respondent has expressed her view on this issue, reply in the local Marathi language as follows:

> Don mule asunahi sambhal karu shkat nahi, vegle rahave lagte, kaslach aadhar nahi, ektepan, dukha, nirasha vatate. Samadhani nahi, niradhar yojnechi pan madat nahi. Aata marnachi vat baghat aahe, jagnyachi ichhach geli marun
> (Even though I have two sons, I am forced to stay alone. They are less bothered about my well being, forget about material, financial and emotional support. I am feeling lonely, hopeless, and sorrowful/regretful. I also don't get any help from the government. Now I don't want to live anymore, how I prefer to die! Or I am waiting to die.)

Table 8.8 shows the percentage distribution of elderly respondents who perceived ageing as a problem by the position they hold as the head of the household.

Table 8.8 Opinion of elderly by their headship

Household head	Being aged is a problem		Total
	Yes	No	
Non-elderly head	64.6	35.4	178
Elderly head	54.6	45.4	405
Household head			
Non-elderly head	64.6	35.4	178
Elderly head in one-member HH	69.5	30.5	59
Elderly head in two-member HH	53.2	46.8	109
Elderly head in more than two-member HH	51.5	48.5	237
Total	57.6 (336)	42.4 (247)	100.0 (583)

Chi-square = 11.495 @ $P < 0.001$
Source Field survey

About 65% of elderly respondents who did not head their households perceived ageing as a problem when compared to those who were the heads of their households (about 50%). Among the elderly who were not the heads of the households, 65% felt that ageing was a problem, while 70% of those who were heads of one-member households viewed ageing as a problem.

The percentage is considerably less among those who were head of households with two or more members (53% for two-member households and 52% for households with more than two members). This clearly indicates that headship of the household is an important factor, which influences the perception of the aged about their status and position in the family.

Residence During Old Age

Awareness of old age homes among the elderly was relatively high (37%). A majority of elderly (86%) reported that the best place for a person to live in old age is with their sons, while the remaining respondents have said that it is with a spouse, alone or with a daughter. Remarkably, 5% of the elderly respondents have reported that an old age home is the best place for a person to live in old age. A possible reason for this response is that their current living environment is not conducive, rather they would prefer, a more social environment.

Responsibility of Elderly Care

The question of who should take care of aged parents becomes critical, especially for those who do not have the resources to meet their own expenses. An overwhelming proportion among the elderly (85%) stated that sons should take care of the elderly (Table 8.9). Only 6% have stated that it was the government's responsibility to help the elderly. Of the elderly, 7% view that they are responsible for their own care and 2% stated that daughters and other members of the family should take care of elderly parents in old age.

Table 8.9 Percentage distribution of elderly according to sex by opinion on responsibilities of care of the elderly

Who should take care of the parents in old age	Sex		Total
	Male	Female	
Self	7.4	7.9	7.6
Sons	84.6	84.5	84.6
Daughter	0.7	1.9	1.2
Government	6.4	4.9	5.7
Others	1.0	0.8	0.9
Total	100.0 (299)	100.0 (265)	564 (100.0)

Source Field survey

Are Children Supportive of Their Parents in Old Age?

It is a common belief in India that children are the main support for parents in old age, which is reflected in the survey with elderly participants reporting 'yes' (83%) and 15% of them 'no' and 2% did not respond. Among the elderly respondents, 21% have said that their view about children being the main support has changed over the years.

> A 78-year-old male respondent has said in Marathi "Aadhi vataiche ki mule hi matarpa-nachi kathi hotel/sahara deti, l pan te sarva khote aahe. Te aaplya bapalach noukara sarkhe vagavtat"
>
> (Initially, I thought my children would take care of me in old age and also give shelter, but I was wrong to think that way. It's shameful to state that, they now treat me like a servant). He further says,
>
> Jasjashi disha badalte tase badaltat sarva lok/mula
> (With change in time, there is change in the behaviour of individuals/children).
> When the elderly were asked "How many sons a person should have to support them in their old age?" most report one or two per person. One interesting reply given in Marathi language:
>
> Mule apekshit aadhar det nahi, tyanchavar avalambun rahaila nako, chagla asel tar ekach pure 4–5 chi aavshakta nahi
>
> (What's the point in having 4–5 children, when they are not helping? Number doesn't matter much. Even one child who is responsible and helpful is better than many.
>
> It is better to have only one child who is helpful and responsible than having many irresponsible sons.)

Summary and Conclusion

The study mainly focuses on how elderly people view ageing and their perception about the younger generation. Due to rapid modernization and urbanization, society is changing in terms of numerous influences on individuals' behaviour. However, one such development is the increasing individualism in today's society across the world. Generally, people think of themselves rather than others. A large number of elderly from the 'oldest of the old' age group (80+) have a negative perception towards the younger generation. Females living alone with a low standard of living also tend to demonstrate a negative perception towards the younger generation. This negative perception among the oldest–old group may be due to the fact that this group needs more care and support from their children, who are unable to provide them with the expected care and support. The elderly who live alone do not have their children to support them or their children may not be willing to keep aged parents with them. To run the house, these elderly living alone were forced to work, which in turn may lead to a negative perception on their part. It is very surprising to see that majority of the elderly respondents perceived ageing as a problem.

The study also shows that the elderly living alone with their spouses develop a more negative perception towards the younger generation as compared to those living with their children. This could be because the elderly feel they have been left alone by their children or relatives to fend for themselves. It also highlights the need for company of either married or unmarried children to make the elderly feel more secure. The elderly in non-headed (children) and single-headed households were more likely to feel that ageing is a problem. It may be because the elderly lose the decision-making power in the family and were forced to ask for minor things from their children, as children were the decision-makers. Children may not discuss family issues with their elderly parents or may share only after the decision has been made, resulting in elderly feeling excluded from playing an important role in the family, which may affect the elderly's emotions and make them feel that getting old is a problem. These factors are commonly considered as barriers in the ageing process, which often result in the elderly feeling that ageing is a problem.

The study highlights the need to support elderly, not only in terms of economic support but also morally and emotionally. These results demand for policies to provide care for elderly's health, living arrangements, pension schemes, etc., thus improving their overall status. In the traditional society, the support system was given by the children and relatives. However, over time, this support system has been deteriorating; thus, there is a need for government to step in and fill this gap and provide support its elderly citizens.

Acknowledgements This paper is based on the author's Ph.D thesis 'Health, Family and Social Adjustment of Elderly in Rural Maharashtra', 2007. At International Institute for Population Science, Mumbai 2007.

References

Asharaf, A. (2000). Elderly in Kerala: Productive units with challenges ahead, unpublished Ph.D. Thesis, *International Institute for Population Sciences*, Mumbai.

Behara, N. K., & Mohanty, R. P. (2005). *Aged in changing social system and their problems*. New Delhi: Discovery Publishing House.

Census of India. (2001). Primary Census Abstract of Maharashtra. Directorate of Census Operations, Maharashtra, Serial, 28.

Gaur, R., & Amarjot, K. (2004). Health status and general attitudes of institutionalized and non-institutionalized aged persons living in and around Chandigarh. *Man in India, 84*(3–4), 153–166.

Radha Devi, D., Santosh, S., Ashraf, A., & Roy, T. K. (1999). *Aged in a changing society: A case study of Kerala*. Mumbai: International Institute for Population Sciences.

Rao, P. A. (2001). From Home to Old Age Home: Aged in Transition in An Urban Setting in Karnataka. Unpublished Ph.D. Thesis, International Institute for Population Sciences, Mumbai.

Chapter 9
Importance of Caregiving in Accomplishment of Daily Routine Activities of the Oldest-Old

Shubham and A.K. Joshi

Abstract Aging is taking place as a broad social trend. This study focuses on how caregiving plays an important role in accomplishing daily routine activities which are essentially followed by the oldest-old. The study outcome is expected to provide a better understanding of the status of oldest-old and bring out the issues that need interventions to ensure their well-being in urban society. The study was a descriptive cum exploratory study done by using case study method with structured and in-depth interviews along with general observations of oldest-old persons and their caregivers separately. A total of five case studies were taken from the urban locality of Kolkata. Results indicated that caregiving plays a pivotal role in the accomplishment of daily routine activities of the oldest-old person, as their dependency on caregivers was found to be high in various aspects. With the social change in family fabrics in urban society, where children are working and no one left to look after the oldest-old persons at home, there is a strong felt need of assisted care, effective social work interventions, providing a helpline for elderly and home-based services.

Keywords Oldest-old · Caregivers · Daily routine activities

Introduction

Aging is a universal process and it affects each human being in the world. It is a consequence of demographic transition, i.e., the change from high fertility and mortality rates to low fertility and mortality rates. One of the major features of demographic transition in the world has been the considerable increase in the

---

Shubham (✉) · A.K. Joshi
Department of Sociology, Banaras Hindu University,
Varanasi 221005, Uttar Pradesh, India
e-mail: shubhambhu08@gmail.com

A.K. Joshi
e-mail: arvindvns@outlook.com

S. Irudaya Rajan and G. Balagopal (eds.), *Elderly Care in India*,
DOI 10.1007/978-981-10-3439-8_9

absolute and relative numbers of elderly people. This has been especially true in the case of developing countries like India. Increase in life span also results in chronic functional disabilities creating a need for assistance among the oldest-old to manage simple chores of life.

Traditionally, in India, the most common form of family structure was the joint family, which consisted of at least two generations living together, and this arrangement was usually to the advantage of the elderly, as they enjoyed a special status and power. But with growing urbanization and depending on the availability of jobs, children are moving out of the joint family setup, leaving the "empty nest" and establishing their own nuclear families.

In the coming years, the elderly population will grow in numbers, and at the same time, the family size will reduce, more so in the urban areas. In the absence of traditional caregivers, due to the disintegration of the joint family and women moving out of the household, the elderly have become a vulnerable group, needing care and attention.

Background

Population aging is the most significant consequence of the process known as demographic transition. Reduction in fertility leads to a decline in the proportion of young in the population. Coupled with fertility decline, reduction in mortality enhances the life span of individuals leading to higher life expectancy at older ages (United Nations 2005). The Indian aged population is currently the second largest in the world after China (100 million). Table 9.1 describes population of selected age groups in India from 1991 to 2011 (in millions). The absolute number of elderly in India is projected to increase from 77 million in 2001 to 137 million in 2021 (United Nations 2005).

The decadal growth rate among elderly population during 1991–2001 is about 40%—double than the general population growth of 21%. The population of the

Table 9.1 Population of selected age groups in India from 1991–2011 (in millions)

Age group (years lbd)	Census 1991*	Census 2001*	Census 2011*
All ages	838.6	1028.6	1210.6
0–4	102.4	110.4	112.8
5–9	111.3	128.3	126.9
10–14	98.7	124.8	132.7
15–59	464.8	585.6	729.9
60–99	56.5	76.5	103.2
100+	0.2	0.1	0.6
Age not stated	4.7	2.7	4.5

**Note* (Excluding Jammu & Kashmir, Excluding Mao Maram, Pao Mata and Purul Sub Divisions of Senapati district of Manipur). *Source* http://www.censusindia.gov.in

world stood at around 6.1 billion in the early twenty-first century and is projected to increase to 9.4 billion in 2050 and 10.4 billion in 2100. If we compare the global population, it has doubled between 1950 and 2000 and is likely to add another 4.4 billion in the next 100 years. The proportion of elderly is expected to grow from 9.9% in 2000 to 14.6% in 2025 and 21.1% in 2050, respectively, (United Nations 2005). Among the elderly, the proportion of oldest-old (80+) is likely to increase from just 1.1% in 2000 to 3.4% in 2050 and 7.1 in 2100. In the beginning of twentieth century, the life expectancy for India was just 23 years for both sexes (Davis 1951).

One of the most remarkable demographic developments in modern times is the progressive demographic aging of the older population itself. In virtually all countries, the most elderly of the older generation (often referred to as the "oldest-old") is growing faster than its younger segment. Indeed, according to United Nations Population Division (2002) projections, the average annual growth rate of persons aged 80 years or over (3.8%) is currently twice as high as the growth rate of the population over 60 years of age. Moreover, the proportion of those older than 80 is projected to increase almost fourfold over the next 50 years to 4.1% in 2050 (United Nations Population Division 2002). Thus, twenty-first century is the century of old (Leibig and Irudaya Rajan 2003), the dependency among aged in India. The number of oldest-old in India was 80,38,718, out of which 20,22,345 live in urban areas (Census of India 2001). Table 9.2 reveals the percentage share of different age groups in India during 2001–2011, which clearly indicates the rise in elderly population including the oldest-old. India has emerged as "aging India" in the beginning of the twenty-first century.

In India, the traditional family setup had been providing social security to the elderly. Studies have shown there is an emergence of social isolation among the aged (Goswami 2005; Irudaya Rajan 1999a). This social isolation affects the living pattern of the aged. The problem will be aggravated in the future as the system undergoes rapid modernization and transformation (NSSO 2006).

Table 9.2 Percentage share of different age groups in India from 2001–2011

	India	
Age group	2001	2011
All ages	100.00	100.00
0–4	10.74	9.32
5–9	12.47	10.48
10–14	12.14	10.96
15–59	56.93	60.29
60–99	7.44	8.53
100+	0.01	0.05
Age not stated	0.27	0.37

Source http://www.censusindia.gov.in

In India, people generally respect the aged and take care of them in a respectful manner. Conventionally, the family system had the primary responsibility of the taking care of the elderly. In most cases, the elderly lived with their son or daughter (Nayar 1992; Desai 1985; HelpAge 1998).

Recent years have witnessed social and economic transformation that has resulted in the disintegration of the joint family system and the rapid decrease in the family size which has made the elderly isolated. A study by Dak and Sharma (1987) highlights a decline in the role of the aged in the family, as they get isolated in urban India. High incidence of migration and urbanization has put the elderly in stress (Irudaya Rajan 2004; Alam 2004). Conditions of the elderly without proper familial support, i.e., living without son or daughter in the house are a sign of lack of social and emotional security among the aged.

Aging is characterized by loss of physical function, social relationships, and cognitive function. In the 1960s, social gerontologists debated two opposing theories of successful aging (Havighurst 1961). The disengagement theory suggested that gradual disengagement from social life and an increasing focus on the inner self is a natural process of human aging, because physical and mental decline is inevitable (Havighurst 1961). Conversely, the activity theory argued that maintaining a middle-aged lifestyle (i.e., remaining active) into old age is important for the elderly. The former theory emphasized passive acceptance of aging, whereas the latter emphasized the active avoidance of aging. Both theories have some merit for explaining successful human aging, but neither has been directly tested. Subsequently, the continuity theory was put forward, advocating a combination of both the disengagement and activity theories, and that continuous selection of either theory based on individual preference is the path to successful aging (Atchley 1989).

The Berlin Aging Study of the oldest-old has reported that they have inferior learning ability, lower life satisfaction, and a decline in positive affect. This state is referred to as "psychological mortality," and helping the oldest-old to maintain their dignity is a large burden for developed countries (Baltes and Smit 2003). Recent studies have added new evidence suggesting that there is a paradoxical relationship between physical health and psychological well-being in centenarians and the oldest-old (Jopp and Rott 2006).

Social loneliness is defined as negative feelings about being alone. Studies from Western countries show that the aged are more socially isolated after retirement and their detachment from work (Keith 1994; Maddox 1999).

The health and well-being of older adults is affected by the level of social activity and mood states. Researchers have reported the negative effects of loneliness on health in old age (Heikkinen et al. 1995). Loneliness, coupled with other physical and mental problems, gives rise to feelings of depression in the elderly persons. Gender differences have been reported in the prevalence of health problems in elderly persons (Arber and Ginn 1991).

Activities of Daily Life

The Index of Independence in Activities of Daily Living (ADL) is based on an evaluation of the functional independence or dependence of patients in bathing, dressing, toileting, transferring, continence, and feeding. Specific definitions of functional independence and dependence appear below the index. Independence means without supervision, direction, or active personal assistance, except as specifically noted below. This is based on actual status and not on ability. A patient who refuses to perform a function is considered as not performing the function, even though he or she is deemed able (Katz and Downs 1970).

Basic ADLs (BADLs) consist of self-care tasks, including bathing and showering, dressing, eating/feeding including chewing and swallowing, functional mobility like moving from one place to another while performing activities, personal hygiene and grooming including brushing/combing/styling hair, toilet hygiene, i.e., completing the act of urinating/defecating (Katz et al. 1963).

Instrumental activities of daily living (IADLs) are not necessary for fundamental functioning, but they let an individual live independently in a community (Roley et al. 2008). It includes housework, taking medications as prescribed, managing money, shopping for groceries or clothing, use of telephone or other form of communication, and using technology and transportation within the community.

The importance of physical health for psychological well-being has been reported in a number of studies. Revicki and Mitchell (1990), for example, found that physical health was the most important source of life strain among older adults. Physical health can have a major impact on subjective well-being. For instance, Bishop et al. (1986) found that poor health was a significant factor in lower morale. Heidrich (1993) examined the relationship between physical health and psychological well-being in elderly women. The results indicated that poor health was associated with more depression and anxiety and lower levels of positive relationships and autonomy in elderly women.

In a longitudinal study of 193 individuals among the oldest-old living in a Swedish municipality, Bravell et al. 2010 describe the last year of life of a sample of the oldest-old, focusing on care trajectories, health, social networks, and function in daily life activities. Most of the elderly in this sample of the oldest-old (74.5%) died at an institution, and the relatives were mostly satisfied with the end-of-life care. The relatives estimated that the health steadily declined during the last year of life, and that there was a decline in performing of daily life activities. They also estimated that those dying in institutions had fewer social contacts than those dying in a hospital or at home. Care at end-of-life for the oldest-old is challenged by problems with progressive declines in ability to perform ADL and health. The findings also highlight the need to support social networks at eldercare institutions (Bravell et al. 2010).

A study by Horgas et al. (1998) describes the daily life of the very old in terms of frequency, duration, variety, and social and physical contexts of activities and to examine the effects of background variables such as age, sex, residential and marital status, income, and education on late life activity engagement. A representative sample of 516 adults aged 70–105 was interviewed about their activities using the Yesterday Interview. The results indicated high frequencies of obligatory activities but also showed substantial time spent in discretionary activities, with television viewing occupying most of the participants' leisure time. Most activities were done alone and at home.

Need for the Study

It was felt that there is a need of one-to-one contact with the oldest-old subjects in order to know their daily routine activities and the problems they face in accomplishing them. Caregiving plays a vital role in the accomplishment of the needs and activities of daily living of the oldest-old. This study focused on the caregiving patterns provided to oldest-old person by their primary caregivers and problems faced in accomplishing their daily routine activities. It was felt that oldest-old persons residing in nuclear families with working children need attention in order to understand their issues related to aging problems, declining health, isolation and care in the urban areas, and their existing support system at the family and community level. The study outcome is expected to provide a better understanding of the status and bring out the issues that need intervention.

Objectives of the Study

I. To assess the daily routine activities and dependency of the oldest-old on their caregivers.
II. To assess the percieved burden of care in the family members of oldest-old (80+ years).
III. To assess the support system and care available within the family and community for the oldest-old.

Research Design

Present study has an exploratory cum descriptive research design. Both quantitative and qualitative research techniques were used. This study was done by using case study method with structured interviews as well as general observation with oldest-old men and women along with their caregivers. The qualitative component included in-depth interviews (IDIs) carried out with the subjects and their caregivers separately. As this study has an exploratory cum descriptive research design, therefore few research questions were set in order to draw meaningful results and reach a conclusion for necessary interventions.

Research Questions

a. How oldest-old persons carry their daily routine and which activities constitute their daily routine?
b. How much dependency they seek from their care providers in accomplishing their daily routine activities?
c. What kind of leisure activities oldest-old enjoy in their routine life?
d. What problems and burden do their caregivers face in providing assistance in the accomplishment of their daily activities?
e. How aging process and physical decline interplay with the psychological and mental states of the oldest-old?

Methodology

Sampling A total of five case studies were selected. Data were collected randomly from an urban housing society of Kolkata. Sampling technique used was purposive random sampling. The sample size was kept small because the needs, ways of caregiving, and pattern of daily routine activities were varied extremely with different subjects due to factors such as individual, familial, and sociocultural differences.

Time period Duration of the study was 3 months conducted between May and July 2014.

Tools Used Consent form with detailed sociodemographic data sheet and a structured schedule with a checklist of activities of daily living (ADL), 2008, (WGBH Educational Foundation and The Massachusetts Institute of Technology).

Procedure Through interviewing the oldest-old and their caregivers, a detailed sociodemographic parameters of all the subjects were studied. The present physical, mental, and psychological assessments of the oldest-old subjects were done. The daily routine activities of the subjects were noted by general observation method by staying along with them and having casual conversations. ADL and IADL were sorted out through an ADL checklist. Feeling of isolation and loneliness was assessed accordingly to the urban life aspects. Leisure time activities of the oldest-old were explored. Adjustment process to gross changes among the oldest-old was studied through objective observations.

Responses of the caregivers of the oldest-old were interviewed separately in order to know the kind of burden or problems they face in providing daily care. The pattern of caregiving and kind of support system available for the oldest-old care according to their daily requirements was noted. Burden of care was assessed in terms of the following domains: physical, financial, recreational, social, and psychological. Practical problems faced by the caregivers in providing care were assessed.

Case Studies

Case A

Case A is a 81-year-old male, residing along with his joint family including his wife, two sons, two daughters-in-law, three grandchildren, a widowed sister-in-law, and his eldest sister. He is the owner of a big flat in a posh urban area of Kolkata. Case A and his family follow Hinduism and belong to a big business class Marwari family whose origin is in Rajasthan but have settled in Kolkata two generations back. They speak Hindi and Rajasthani dialect. Case A is qualified up to matriculation, and as he got absorbed in his family business, education was not given much priority especially in business class families. Case A looks after a family business of plastic materials, and he worked very hard to establish this business and is working for last 60 years. He currently maintains records and accounts of his family business which is now handed over to his sons. He described his life which was full of struggles initially because he had to run a big family and support his father in establishing the new business. He started working at the early age of 15–16 years with his father. They expanded their business from a small scale to big. Meanwhile, he got married at the age of 22 years and had two sons and a daughter. He and his wife being the eldest in family looked after their parents and children and all the responsibilities of a joint family including his younger brother who stayed with them with his family till the business was partitioned between the two. After the death of his father, he took all the family responsibilities of taking care of his mother and other members. After the death of his brother, his widowed wife also started staying with them. The family bonds between the members were strong and very stable, which has helped him to continue things well as he grew old. Case A confidently assures that for economic needs he was always independent. In all the domains of ADL checklist, Case A was found independent and satisfactorily well maintained. His leisure time activities include watching television, performing religious tasks, and socializing with his community people. Case A perceived his quality of life as satisfactory and described his subjective feelings of isolation and loneliness as very low because he had good family environment with enough support from the members of the family and was socially recognized in his community.

Chief caregivers of the subject were his wife, sons, daughters-in-law, and grandchildren, where sons are working and daughters-in-law are homemakers. His wife and daughters-in-law were interviewed separately and they described that he follows a disciplined lifestyle, is very strict about his routine, and carried a positive attitude toward life. He is a lively and fun-loving person and leads an active social life. He always maintained a good physical status, orientation, cognition, memory, and self-care. For physical exercise, he used to walk daily and performed all his daily household routine tasks on his own such as bathing, toilet, eating, cleaning room, washing clothes, maintaining hygiene, taking medication, walking, climbing stairs, going to park, following news, use of phones, watching television, and going to nearby places for shopping by using public transportation. Subject's wife described him as flexible to new adjustments. Caregivers did not feel any burden in providing care for Case A, as he was physically and economically independent for his maintenance and was actively participating in daily chores of life. Thus, no significant problems were found in caregiving and perceived burden on the family.

Case B

Case B, who was a 80-year-old married, male, resided in a joint family in an urban area of Kolkata. He followed Hindi dialect and is qualified up to postgraduation. Currently, he is a retired government servant and has upper-class socioeconomic status. Case B and his family follow Hinduism. Case B stated that he belonged to a Brahmin family of Varanasi who settled down at Kolkata after getting a job. His father was a small businessman who stayed at their village with his mother and siblings. He worked in a jute mill for about 40 years. He is retired for last 20 years and is currently staying at home with his wife, elder son, daughter-in-law, and two grandchildren. He has three children, which includes two sons and one daughter. His sons are settled well, and the daughter who is married stays at her in-laws house. He described that post retirement he found himself to be isolated and depressed because he had unsatisfactory job, which did not help him acquire things which he desired in life. Because he had to earn for his family, he worked at the jute mill which had hectic duty hours and bad working environment. For economic needs, Case B is dependent on his pension and other requirements are fulfilled by his son and daughter-in-law. Subject's view about his life as a whole was a mix of ups and downs which he shared. He showed happiness about his children's achievements. Case B described poor perceived quality of life, as he found his life dissatisfactory and one of his underachievement. His leisure time activities included watching television, walking inside the community park, or reading newspapers. He did not do much physical activity, but used to perform all his ADL, religious ritual, eating, maintaining hygiene, taking medication, walking, climbing stairs, going to park, following news, use of phones, watching television, and going to nearby places for shopping by using public transportation on his own. Case B described his subjective feeling of isolation and loneliness as high because he is not socially active and has poor confidence in interacting with people. There is feeling of inferiority due to the hearing loss, which made the subject to remain aloof and isolated. He said that even during his younger days, he had poor communication skills which made him reserved in nature. The leisure time of Case B was not utilized well.

Chief caregivers of the subject were his wife, son, and daughter-in-law. His wife is a homemaker, son is a doctor, and daughter-in law is a government servant. Subject's wife and daughter-in-law were interviewed separately who described that Case B has poor physical health with moderate level of hearing impairment and heart disease. He had a thin body but had proper orientation, cognition, memory, and self-care. His sleeping and eating pattern was adequate. His routine schedule was properly followed and physical exercise included walking. According to the caregivers, Subject B did not participate much in daily household tasks and was dependent on his caregivers for the accomplishments of certain tasks. He has limited social life. It was observed that loss of quality time for the subject was significant, as both caregivers were working. Overall, perceived burden on family regarding caregiving was moderately high due to subject's poor health and isolated behavior.

Case C

Case C is a 81-year-old widow, Hindu Punjabi, female, residing along with her family including her daughter and son-in-law in urban area of Kolkata. They followed Hindi and Punjabi dialect. Case C was qualified up to class 8th and was a homemaker. Socioeconomic status of the subject's family is upper class. For economic needs, subject C is dependent on her son-in-law as her daughter is also a homemaker. Case C during the interview said that she was a native of Punjab where she lived with her family and got married at the tender age of 17 years, as early marriage was the norm those days. After discontinuing her school education, she stayed at home with her parents and learnt household chores in preparation for a married life. She had an arranged marriage. After marriage, she stayed with her husband and in-laws and she had a huge joint family in Punjab. She gave birth to her only daughter at the age of 22 years. Her husband was a businessman who had a garments store at Ludhiana. Her husband died at an early age of 51 leaving her behind family responsibilities and the garments shop. She felt all alone after the death of her spouse, which led to economic problems, depression, and anxiety. Later, her daughter brought Case C to her house, selling all the properties at Punjab, and has stayed together for the last 18 years in Kolkata. She misses her granddaughter a lot who stays at USA. Case C described her poor health and deteriorating physical activities due to locomotor problems in joints of the lower limbs causing compromised movements. However, she described that she is emotionally and psychologically stable and happy with her current life. She appreciates the efforts taken by her daughter and son-in-law with her caregiving. She willingly participates in daily household routine tasks such as involvement in kitchen work and cleaning, which required less physical efforts. Her leisure time activities included watching television or performing religious rituals. According to her, perceived quality of life was satisfactory. She followed her schedule with assistance of her daughter in accomplishing daily routine activities as per the ADL checklist. Subjective feeling of isolation and loneliness as revealed by Case C was found low as she had harmonious relations and good interactions with the family and community.

Chief caregivers of the subject were her daughter and son-in-law, where son-in-law is the earning member. Both of them were interviewed separately and they said that subject C followed daily routine activities on her own such as bathing, toilet, eating, cleaning room, maintaining hygiene, taking medication, reading newspapers, using phones, and watching television. She suffered from locomotor disability in her lower limbs for last 3 years because of which she walked with the help of a stick or support, and she could not climb stairs or go to park alone. She had an average body built with good orientation, cognition, memory, and self-care. Sleeping and eating pattern was adequate, but physical exercise was absent. According to the caregivers, Case C was flexible to new adjustments. Subject C was dependent on her caregivers for going out of house, for shopping. and moving to nearby places; therefore, perceived burden of caregivers was found moderately high.

Case D

Case D is a 82-year-old widow, Hindu, male, residing along with his nuclear family in an urban area of Kolkata, following Hindi and Bhojpuri dialect, qualified up to secondary school and is currently a retired businessman. He was owner of a general store in a small village in Bihar. He started working as a small businessman at the age of 18 years. His father owned a small agricultural land and a mango garden at the village. He also had a joint property with his elder brother where they stayed together. He got married at the age of 19 and had four kids. His sons and daughters received education and migrated to cities for higher studies and jobs. He worked hard to earn a better livelihood for the family. Later, he sold the agricultural lands for marriage of his daughter and building of his own house. His wife died suddenly of a heart attack at the age of 45 years, which led the whole family into a sate of shock. Somehow, they recovered from the trauma, but the setback of being alone became a concern for his family with growing age. There was a pressure from his children to migrate to the city and stay with them. Case D was quite reluctant on the issue of migration to his son's house, which caused stress. Leaving his business and house and staying without work at his son's house was not acceptable for him. He did not want to be dependent on others and lose his authority. After the death of his elder brother and on completing 75 years, Case D became weak and he decided to wrap up the business and went to stay with his eldest son and daughter-in-law at Kolkata. He is currently not working but earns his livelihood from the rented house at his village. Socioeconomic status of the subject is upper class. For economic needs, subject D is independent. He has a good physical status with an average body built but had moderate level of hearing impairment. He possessed good orientation, cognition, memory, and self-care. Sleeping and eating pattern was adequate with well-maintained hygiene. Physical exercise in form of walking was done. In most of the domains of ADL, Case D was independent and well maintained. Subjective feeling of isolation and loneliness was found to be average, as the subject had compromised communication pattern due to hearing loss and has reduced social life in comparison with earlier life. Leisure time activities were only restricted to newspaper reading, watching television, visiting temple, and doing religious rituals.

Chief caregivers of Case D were his son, daughter-in-law, and a servant, because both the primary caregivers are doctors and are not available at home always. His caregivers revealed that he was found to be physically active in daily routine activities, as he used to perform all his ADL on his own such as bathing, toilet, eating, maintaining hygiene, walking, climbing stairs, going to park, following news, use of phones, watching television, managing accounts, shopping, and going to nearby places by using public transportation. Time schedule in daily routine is properly followed. He shared congenial social relations with family relatives and very few friends and often visits his younger son's house and his native village alone. No significant problems were found in caregiving pattern, and perceived burden on family regarding caregiving was low. Caregiving was shared among the other siblings also.

Case E

Case E was the eldest among all the cases of oldest-old under study. He was a 92-year-old married, male, residing along with his wife, son, daughter-in-law, and a grandson in an urban area of Kolkata. The native place where the case was rooted was Varanasi where he was born and brought up in a Brahmin family. He studied and worked at Varanasi where he served in government service for 38 years. He followed Hindi and Bhojpuri dialect and was qualified up to postgraduation. He got married at the age of 24 years and had four children. They stayed in a joint family until his children grew up and got married. After completing the education, his children migrated to various places for jobs. Subject E resided in Varanasi at his own house with his wife till his retirement. Soon after retirement, Case E started staying with his younger son and daughter-in-law. Due to some economic and family problems, he came to stay with his elder son, daughter-in-law, and a grandson at Kolkata. He is currently retired and not working for last 32 years. Socioeconomic status of the subject is upper class, but for economic needs, subject E is dependent on his pension, rented property at native place, and his caregivers. Subject had a poor physical status with a thin body build and had moderate level visual and hearing impairment with locomotor disability, with poor orientation, average cognition, and loss of memory with poorly maintained self-care. Physical exercise was completely absent. Mostly, leisure time activities included watching television and reading newspapers and magazines. Perceived quality of life was found satisfactory. In most of the domains of ADL, Case E was found dependent on caregivers and servants. Subject E was not found to be physically active, as he had marked locomotor dysfunction and age-related severe disability. Social life of the subject was found to be highly compromised for last few years as compared to earlier life due to loss of functional physical ability. Perceived feeling of isolation and loneliness was found moderate as subject was found to have a good family support. Leisure time activity remained lacking.

Chief caregivers of the subject were his wife, son, daughter-in-law, and servants, where son and daughter-in-law both are working and spent less time at home. Caregivers were interviewed where they described the kind of dependency Case E was having on them to assist him in performing all his routine tasks such as bathing, toilet, eating, cleaning room, maintaining hygiene, taking medications, walking, climbing stairs, going to park, following news, use of phones, watching television, managing accounts, and shopping. Time schedule in daily routine is properly followed with help of maids. Sometimes, due to unavailability of the servants, the situation became complicated because the subject became anxious and irritable. His rigidity in nature also caused problems for the caregiver as Case E sometimes refuses to take food being provided, did not bathe to avoid falls in the bathroom, and did not cooperate in change of clothes, diapers, etc. There were problems like lack of quality time for the subject by the caregivers. He shared congenial social relations with family relatives who often visited him on occasions. Overall, subjective burden of care of the subject on the caregivers was reported high.

Findings and Observation of Caregivers

The basic concern of all the caregivers under this study was that, with such an advanced age, there is always a fear of some wrong happening and insecurities regarding the care of the oldest-old cases. In urban life where nuclear family resides in flats and males and females are working, social life becomes very formal and restricted, which is not found in village life and joint families. This is also obvious that lack of joint family increases the burden of care which can be shared with other members too. In urban areas, there is a lack of assisted care, palliative care, geriatric nursing, and day care centers for elderly persons, where they can be taken care of in the absence of the working caregiver. There is a substantial lack of care of the aged in the community, as there are no pathways, lifts, ramps, vehicles, or any other special provisions for physically disabled elderly or exclusively elderly people. This is a reason why caregivers do not allow the elderly to go out of house alone. This restricts the social life of elderly adversely, leading to more loneliness and isolation. These are serious concerns which have made caregivers stressed about their responsibilities to their parents due to lack of time, finances, availability of resources, etc. It has markedly affected mental and physical health of the caregivers and reduced their social life as well. They cannot leave their aged parents alone at home and go anywhere for a long time. There are several cases of violence and theft with elderly person resulting in fatal accidents. It is very difficult for caregivers to handle the physical and mental health of their elderly parents along with their own busy life, as they are not always available around them for emergency situations.

Conclusion

Findings indicate that caregiving plays a pivotal role in the accomplishment of daily routine activities of the oldest-old persons, as their dependency on caregivers was found significant in all the cases. In the urban society where both men and women are working, the care of oldest-old becomes compromised. With the social change in family fabrics in urban society, where children are working and no one is left to look after the oldest-old persons at home during their absence, there is a strong felt need of assisted care and effective social work interventions providing home-based services such as health care, assistance in completion of daily routine activities, utilization of leisure time, physiotherapy, and counseling.

In all the cases, it was marked that oldest-old persons complained about not being provided with quality time by their caregivers when needed due to their busy life schedule and work. All the respondents were found to have satisfactory help, attention, and care.

From the above case studies, it can be concluded that the major problems faced by oldest-old persons are gross physical decline in health-related problems, locomotor disabilities, and marked impairments in vision and hearing causing discomforts and complexes related to aging process.

Subjective feelings of loneliness and isolation were found to be moderate to high in all the subjects depending upon the type of social interaction and family support available for the oldest-old. Subjects who were widowed were found to have more loneliness in comparison with those who had their spouse. All the cases lived with their family and were looked after by their children or their spouses.

All the respondents belonged to an upper class socioeconomic status in urban areas. Dependency of the oldest-old on their caregivers for the accomplishment of their daily routine tasks was found moderately high. Financial burden was not significant, as all cases belong to upper class and they possess all kinds of resources and help for the care of the oldest-old. No significant problems were found in caregiving pattern.

Four out of five cases were found to be physically active in performing their ADL, taking medication, walking, climbing stairs, going to park, following news, use of phones, television, managing accounts, shopping, going to nearby places by using public transportation, etc.

The gross disability increased among the oldest-old with ascending age, causing poor health, compromised physical activities, locomotor disability, depression, decreased social interactions, isolation and loneliness which increased their dependence on caregivers.

Utilization of leisure time was markedly lacking. Mostly, all the cases spent their time watching television or walking on the terrace, sitting into balconies, going to the park or talking to any family member. It was found that leisure time activities lacked productivity and enthusiasm.

In urban life where nuclear family resides in flats and where children are working, social life becomes very formal and restricted, which is not found in villages and joint families. This makes oldest-old persons more vulnerable to social isolation and loneliness. This has also affected the mental health of oldest-old, as they are treated as outdated which involves a feeling of inferiority in them. They are unable to exercise their freedom and authority, as children take up the functional responsibilities in the family.

Major Concerns of the Caregivers

Changing social fabric in India has resulted in breakdown of traditional joint family and caregiving support systems among the elderly. There are limitations on independent functioning and transport for elderly persons, as there is lack of ramps, wheelchairs, special bathroom facilities, transportation, pavements, day care centers, old age clubs, health centers, etc. As a result, the elderly in Indian cities is often housebound, uprooted from their native places, and becomes dependent on their children. It is important for the caregivers to make oldest-old feel secure in the home environment by keeping them occupied in everyday task of living, schedule of daily activities, regular exercise, proper communication, comfort and emotional support. Certain activities such as planning vacations, holidays, availability of expert for crisis intervention, attention to physical and mental health, maintenance of social network can help a lot more in providing affective care to the oldest-old. Elderly can be emotionally and sensibly cared by the caregivers. In India, the respect given to the elderly is a positive factor along with the strong family ties. It is often the Dada/Dadi, Nana/Nani who passes down moral education, cultural myths, legends, religious stories, etc. to the grandchildren. This gives them a sense of worth and dignity.

Social engagement includes social connectedness and participation in productive activities. Social support provides opportunity to give and receive support and reduces loneliness. Social structure facilitates purposeful roles, connectedness to peers, and contribution to the development of younger members. Productivity in oldest-old can be enhanced by volunteerism for altruistic reasons, venturing into new pursuits, and creativity gives purpose in life, chance to interact with like-minded peers, and sense of competence. Life satisfaction is a sense of personal well-being determined by feeling of having control and choice, perception of one's situation, adequacy of social support, income and social comparison.

Psychosocial care can play a pivotal role in prolonging productivity and improving quality of life of the elderly. For many years, old age was conceptualized exclusively in terms of decline and loss, but changing perspectives in later years consider it a period of tremendous individual variability, change, growth rather than universal decline. To render elderly care without allowing seniors to lose their sense of dignity and independence is not easy. Therefore, factors contributing to good physical and mental health in the elderly are strong support groups, being socially involved, fulfilling multiple social roles, sense of social worth, independent source of income, religious beliefs and spirituality, active lifestyle and physical fitness (as far as possible), good self-esteem, and personal efficacy.

Social Work Interventions and Recommendations

1) Door-to-door counseling services can be provided to the oldest-old in order to overcome their psychological and mental distress.
2) Indoor recreational activities could be planned for a better utilization of their leisure time as well as physical and psychological well-being.
3) Assisted nursing and care can be provided to the oldest-old with gross physical and locomotor dysfunction in the absence of the caregivers.
4) A time bound combined activities of play and non-formal education can be framed between the oldest-old and children residing in the community for a

better utilization of time and also for combating feeling of loneliness and isolation.

5) Group therapy can also provide greater help in solving problems of oldest-old and their caregivers. It can provide opportunities for oldest-old to share their feelings and engage themselves in planned group activities and discussions.
6) Community awareness, education, and sensitization programs can provide a better and advanced care to the oldest-old. A proper utilization of the community resources, policy making, promoting health insurance and pension schemes, caregiver support, professional training in geriatrics and gerontology, a holistic approach and concerted efforts by a "multidisciplinary team" can provide care and help to oldest-old to a greater extent.
7) Special provisions for elderly with disabilities should be made in communities for their safety and convenience such as assistance with locomotory equipments, ramps, wheelchairs, and brackets.
8) An "Oldies Helpline" can be generated with the provisions of various service delivery systems around the clock, keeping in view the needs and requirements of oldest-old persons. It can deliver immediate help to the oldest-old by extending good care facilities exclusively for them.

Limitations of the Study

Since this study was done on a very small group of oldest-old, generalizations on broader aspects cannot be drawn. As the universe was very diverse, very small sample was selected to avoid ambiguity. Study was limited to upper-class urban society, so less problems and burden were found in caregiving pattern. No scale was used to asses the percieved burden. Time period of this study was also limited.

References

Alam, M. (2004). Ageing, old age income security and reforms: An exploration of indian situation. Economic and Political Weekly, 3731–3740.

Arber, S., & Ginn, J. (1991). *Gender and later life*. London: Sage.

Atchley, R. C. (1989). A continuity theory of normal aging. *Gerontologist, 29,* 183–190.

Baltes, P. B., & Smith, J. (2003). New frontiers in the future of aging: From successful aging of the young old to the dilemmas of the fourth age. *Gerontology, 49,* 123–135.

Bishop, D. S., Epstein, N. B., Kietner, G. I., Miller, I. W., & Srinivasan, S. V. (1986). Stroke—Morale, family functioning, health-status, and functional capacity. *Archives of Physical Medicine and Rehabilitation, 67*(2), 84–87.

Bravell, M. E., Malmberg, B., Berg, S. (2010). E*nd-of-life care in the oldest old, 8*(3), 335–344.

Census Report. (2001). Ministry of Home Affairs, Government of India. http://censusindia.gov.in.

Dak, T. M., & Sharma, M. L. (1987). Changing status of the aged in north indian villages. In M. L. Sharma & T. M. Dak (Eds.)., Aging in India (pp. 43–55). New Delhi: Ajanta Publications.

Davis, K. (1951). The population of India and Pakistan. Princeton University Press.
Desai, K. G. (1985). Situation of the aged in India. *Journal of Indian Anthropological Society, 20*(3).
Economic & health survey on India's oldest old (80+): Need care and access, Helpage India, www.helpageindia.org.
Goswami, I. (2005). *Ageing India, sociological issues*. Presented in Indian Sociological Congress.
Havighurst, R. J. (1961). Successful aging. *Gerontologist, 1*(8–13), 9.
Heidrich, S. M. (1993). The relationship between physical health and psychological well-being in elderly women: A developmental perspective. *Research in Nursing & Health, 16*(2), 123–130.
Heikkinen, R., Berg, S., & Avland, K. (1995). Depressive symptoms in late life. *Journal of Cross Cultural Gerontology, 10,* 315–330.
Helpage. (1998). *Age India, directory of old age homes in India*. New Delhi: Research and Development Division. Helpage India.
Horgas, A. L., Wilms, H. U., Baltes, M. M. (1998). *Daily life in very old age: Everyday activities as expression of successful living*, *38*(5), 556–568.
Jopp, D., Rott, C. (2006). *Adaptation in very old age: Exploring the role of resources, beliefs, and attitudes for centenarians' happiness*; *21*(266–280).
Katz, S., Downs, T. D., Cash, H. R., & Grotz, R. C. (1970). Progress in development of the index of ADL. *Gerontologist, 10*(1), 20–30.
Katz, S., Ford, A. B., Moskowitz, R. W., Jackson, B. A., Jaffe, M. W. (1963). Studies of illness in the aged. *The Index of ADL: A Standardized Measure of Biological and Psychosocial Function, 185*(12):914–919.
Keith, J., et al. (1994). *The aging experience*. London: Sage Publications.
Liebig, P., & Rajan, S. I. (Eds.). (2003). *An aging India: Perspectives, prospects and policies*. New York: Haworth Press.
Maddox, G. (Ed.). (1999). *Encyclopedia of ageing*. New York: Springer.
National Sample Survey Organisation (NSSO, 2006). Morbidity, health care and the condition of aged. New Delhi: Government of India.
Nayar, P. K. B. (1992). Needs and problems of the aged: A case study of Kerala, ICSSR.
Rajan, S. I. (2004). Chronic poverty among the elderly. Working Paper 17, Chronic Poverty Research Centre, IIPA.
Rajan, S. I. et al. (1999a). India National Ageing trends, life long preparation for old age. *Asia and the pacific Journal United Nations*.
Revicki, D. A., & Mitchell, J. P. (1990). Strain, social support, and mental-health in rural elderly individuals. *Journal of Gerontology, 45*(6), S267–S274.
Roley, S. S., DeLany, J. V., Barrows, C. J., et al. (2008). Occupational therapy practice framework: Domain & practice, 2nd edition. *American Journal of Occupational Therapy, 62* (6), 625–683.
United Nations. (2005). United nations department of economic and social affairs, population division. World Population Prospects. New York: The 2004 Revision.
United Nations Population Division. (2002). World population ageing 1950–2050. *Population and Development Review, 28*(4), 814–815.

Chapter 10
Vulnerability and Coping Mechanism of Aged: A Study of Elderly Widows in Jharkhand

Sangeeta Kumari Gupta and T.V. Sekher

Abstract Widows are the most vulnerable segment among the elderly population in India. About 50% of elderly women (60 years and above) in India are widows compared to only 15% among elderly men. Elderly widows suffer multiple problems attributable to gender, widowhood and old age. Based on a sample survey of 300 rural elderly widows from Jharkhand state, this study examines vulnerability and coping mechanisms, using both quantitative and qualitative data. The vulnerability of elderly widows has been measured on the basis of their social relationships, change in status after becoming widow, decision-making ability, etc. "Self–Adjustment Inventory" and "Social–Adjustment Inventory" have been employed to understand the coping strategies. The study has shown that elderly widows face discrimination within their houses and within the community. They have very limited or no role in family decision-making and are economically dependent on others. Lack of property and savings, loss of husband, and poor health make them vulnerable to abuse. The family as a fundamental unit of society needs to be strengthened along with social security measures to take care of elderly widows.

Keywords Elderly widows · Vulnerability · Coping mechanism · Dependency · India

---

The paper entitled "Vulnerability and Coping Mechanism of Aged: A Study of Elderly Widows in Jharkhand" is based on Sangeeta Kumari's Ph.D thesis entitled "Elderly Widows: A Socio-economic and Demographic Study in Jharkhand", awarded by International Institute for Population Sciences (IIPS) in the Year of 2013.

S.K. Gupta
International Institute for Population Sciences, Longitudinal Ageing Study in India (LASI), Govandi Station Road, Deonar, Mumbai 400088, Maharashtra, India
e-mail: sangeetasargam@gmail.com

T.V. Sekher (✉)
International Institute for Population Sciences, Department of Population Policies & Programs, Govandi Station Road, Deonar, Mumbai 400088, Maharashtra, India
e-mail: tvsekher@gmail.com

S. Irudaya Rajan and G. Balagopal (eds.), *Elderly Care in India*,
DOI 10.1007/978-981-10-3439-8_10

Feminization of Ageing in India

Population ageing has been one of the most distinctive phenomena of the twentieth century, and will remain an important issue throughout the twenty-first century. Improved life expectancy has given rise to an increase in the number of persons aged 60 years and above. In 1950, the proportion of elder persons in the world was around 8% and it has increased to 10% in 2000 and projected to reach 21% by 2050. It is estimated that the elderly population will reach almost 1.2 billion by 2025 and 2 billion by 2050 (United Nations 2002).

Population ageing is a world-wide phenomenon. India is no exception to this global phenomenon of ageing with decline in fertility and mortality levels. The Indian aged population is currently the second largest in the world. According to Census of India (2011), the proportion of elderly (60 years and above) in India's population is 8.6% and is likely to increase at much faster pace in coming decades. The aged population in India experiences gradual "feminization" and serious gender disparities. The feminization of ageing is evident from the growth pattern of the elderly females, which remained varying over the past decades (Alam 2006). A major concern relates to the increasing proportion of elderly women, especially widows, in the population. The proportion of widows is much higher as compared to the widowers, a unique characteristic feature of India's elderly. According to Census of India 2011, there were 42 million widows in India constituting 7% of the total female population and 49% of the women over the age of 60 years. Table 10.1 indicates that the age-specific prevalence of widowhood among men is low compared to that of women.

The proportion of widows in the female population rises sharply with age, and it is 34.7% among women in the age group 60–64 years and it reaches 61% among women aged 70 years and above. In every age group, women experience higher widowhood than men, and in the middle age group, nearly thrice as that of men.

Table 10.1 India's aged population by age, sex and marital status, 2011

Aged	Sex	Marital status (proportion)			
		NM	M	W	D/S
60+ years	Males	2.9	82.1	14.6	0.4
	Females	2.0	49.6	47.8	0.5
60–64	Males	2.1	88.8	8.7	0.4
	Females	1.6	63.0	34.7	0.8
65–69	Males	2.8	84.9	11.9	0.4
	Females	2.0	54.4	43.0	0.6
70–79	Males	2.2	78.3	19.1	0.4
	Females	1.6	37.1	60.8	0.4
80+ years	Males	7.5	62.0	30.1	0.4
	Females	4.5	26.1	69.0	0.4

Note NM: Never Married, M: Married, W: Widowed, D/S: Divorced/Separated

Source Census of India 2011

In India, depending upon the marital status of elderly, especially that of women, their position and status within the family as well as in the society will be determined. For many women, "*Widowhood is more than the loss of a husband—it may mean the loss of a separate identity*" (UNFPA 1998, p. 42).

There are three reasons for this unusual gender disparity in widowhood in India —the longer lifespan of women than that of men, the cultural practice of Indian men marrying women younger than themselves, and the higher rate of re-marriage among widowed men compared to the widowed women. Widowhood is characterized as one of the most distressing of all life events (Holmes and Rahe 1967). Mostly, widowhood implies a shift in position from that of a wife with economic, social and emotional security to that of socially and financially insecure women. In spite of the socio-economic changes in Indian society over the decades, the position of widows has not improved much. Widows have to face many forms of deprivation—economic, social, cultural and emotional. The widowed condition places women in a more disadvantaged position in old age and they live as dependents for rest of their life. Elder widows are doubly affected due to the combined effects of old age and widowhood. It was found that elderly females, particularly widows, are more vulnerable to elder abuse and neglect in Indian families (Sebastian and Sekher 2011; Sebastian 2011).

A study by Kumari and Sekher (2010) found that more than half of the elderly were economically dependent on others for their day-to-day requirements, and the situation is worse among elderly females, as widowhood and dependency increases with advancing age. Every phase of widowhood is hard, challenging and deranging and also brings severe social, economic, emotional and cultural deprivation than the widowhood itself (Kumari and Sekher 2012). Though there are many studies on elderly in India, very few focus on elderly widows. There are no statistics which indicate the extent and depth of the vulnerability among the India's elderly widows (Gupta and Sankar 2002).

In India, the implications of ageing and our understanding of this process and its impact are still limited. Most of the studies were based on homogeneous group of elderly people, by overlooking the disparities in the experiences in ageing for men and women. Implications of ageing on the lives of elderly are often taken as natural but there are significant differences in the way ageing affects men and women. A major concern relates to the increasing proportion of elderly women, especially widows, in India's population and their vulnerability. The death of a spouse in old age is known to have profound effects on the well-being of the widow/widower and is an event which provokes important life changes (Momtaz et al. 2009). Indian women who are widowed are widely discriminated against and seldom receive respect and care from people in general, and from the family in particular. India is perhaps the only country where widowhood, in addition to being a personal status, exists as a social institution. Widow's deprivation and stigmatization are exacerbated by rituals and religious symbolism. Indian society, similar to all patriarchal societies, confers social status on a woman through a man. Hence, in the absence of a man, she herself becomes a non-entity, ultimately suffering a social death.

The problems that elderly women face are moreover frequently compounded by their difficulties in obtaining sufficient income because of their limited rights to property. Special attention is needed to provide support and care for elderly widows. However, there is a serious dearth of data on social and economic conditions of widows and their health-related issues (Chen and Dreze 1992). Very few studies have focused on widows as a distinct social group and practically none have attempted to provide a comprehensive analysis of their social and economic conditions. However, widowhood has been an explicit or implicit theme in a number of studies dealing with issues such as old age, living arrangements of the elderly, female-headed households, and remarriage practices, etc.

Widowhood is a turning point in a woman's life, involving new social adjustments for the family as well as for her. Existence of stigmatized social perception, negative attitude towards widowhood and lack of social support systems would manifest in poor mental health and problems of adjustment which will make them more vulnerable. Combined effects of poverty and widowhood aggravate the miseries of women in rural areas. Consequently, they experience more social, economic, health and psychological problems in later years of life. The loss of a companion also leads to weakening of families and social networks, loss of economic power and loss of identity. So, these individuals constitute a special group in our population because of the deprivation, vulnerability and low social status. This paper is an attempt to analyse the vulnerability among the elderly widows, with particular attention to understand the adjustment and coping mechanism adopted by them.

Data Sources and Methodology

The data includes both quantitative and qualitative components. In the quantitative component, primary data collected from selected villages has been analysed. The qualitative information (case studies and key informant interviews) was used to explore issues related to the dependency levels of the widows, relationship with children and others, experience of ill-treatment/abuse, and the extent and nature of deprivation, etc.

Jharkhand state was purposely selected for this study. All districts of Jharkhand were ranked according to the proportion of elderly widows. The district having higher proportion of elderly widows, East Singhbhum (7.8%), was selected for the field study. From the selected district, two blocks were randomly selected. It was targeted to interview 300 respondents from two blocks, with equal number of respondents from each block. In order to complete the targeted sample size (300 elderly widows), 12 villages from the two blocks were visited. In the first block, there are 27 panchayats. One panchayat which was having large number of villages (20 villages) was selected. Before starting the survey, preliminary visit to the study area was made. Out of these 20 villages, 7 villages were covered in the survey. In the second block, there are 14 panchayats. One panchayat was chosen for conducting the survey. From here, 5 villages were covered to collect information from elderly widows. With the help of sarpanch, Anganwadi workers and other locals from the villages, the households with eligible respondents were identified and interviewed using a structured interview schedule. The survey was conducted from March to June 2010.

Profile of the Surveyed Households and Elderly Widows

Distribution by religion suggests that 80% of the surveyed households were Hindus, 13% belonged to the *Sarna*[1] religion and 7% were Muslims. 40% of households were scheduled tribes, followed by other backward castes (37%). The households had on an average 4.7 members. Ninety six percent of surveyed households had one elderly person and around 3% of households had two elderly persons. More than half of the households were female headed. Majority of the households owned no land (55%). Slightly less than one-fourth (24%) of the households had large holdings (land size greater than 3.2 acres). Twenty three percent of the households owning land reported no income from the land. Thirty seven percent of the households reported average yearly income from the land as about Rs. 3500.

Majority of households (94%) owned the structure in which they resided. In case of the housing type, about three-fifths of all households (60%) lived in *kachcha* houses (constructed from mud or other low-quality materials), 16% lived in *semi-pucca* houses and 24% lived in *pucca* houses. More than half of the residential structures contained 2–3 rooms whereas 28% of the residential structures consisted of single room. Only 40% of the households had separate kitchen for cooking, around 30% of the households cook inside their living rooms and 30% of the households cook in open areas. Eighty nine percent of all surveyed households were using dry leaf/wood/coal/charcoal/dung cakes as main source of cooking fuel. Around 8% of the households were using liquid petroleum gas (LPG) and 3% were using kerosene. Seventy percent of the households did not have access to a toilet facility. Around one-fourth of households used kerosene as a main source of lighting. More than half of households (57%) reported that their main source of drinking water was public piped water/hand pump/covered well.

---

[1]Note "Sarna": According to Troisi (2000), Sarna is the Munda word for "Sacred Grove" while Dhorom is the Oriya word meaning "religion". Sarna involves belief in a great spirit called the Sing Bonga. Santhal belief holds the world to be inhabited by numerous spiritual beings of different kinds. Santhals consider themselves as living and doing everything in close association with these spirits. Rituals are performed under groves of Sal trees called Jaher (or sacred grove), where Bonga is believed to appear or express himself. Often, Jaher are found in the forests.

Households/Elderly Benefited from Government Schemes

To understand whether the households had benefited from any government programmes, the information was gathered about the type of government programmes availed by the households. Half of the surveyed households had below poverty line (BPL) cards and received food grains and kerosene meant for BPL card holders. Among the surveyed elderly, around half of them were not receiving pension. Among those who were receiving pension, majority were receiving widow pension (74%) and 19% were getting old-age pension.

The amount of widow pension and old-age pension is Rs. 200 in a month in Jharkhand. All the elderly widows were interested to avail pension from the government, but half of them not getting it. When asked about the reasons for not availing pension, they stated many reasons—"they do not know how to apply, officers told them to wait or they are not eligible for the pension". Few households benefited from the Indira Awaas Yojana (free housing provided by the government).

Another dimension of economic status of households is their borrowing/indebtedness. Twenty-two per cent of the households had borrowed money during the last one year. The most important reasons mentioned were "treating for health problems" (38%), followed by "meeting the household consumption" (23%). Twelve percent of households have borrowed money "to meet marriage-related expenses". Major source of borrowing was from neighbours (39%) and relatives (23%). Only 15% had taken loans from banks and 14% from local money lenders. If they borrow money from relatives or neighbours, they need not pay any interest. Money lenders charge higher interest rate than banks. Thirty per cent of the households had indebtedness of around Rs. 2000, whereas slightly less than one-fifth of the households had indebtedness of more than Rs. 8500.

Profile of the Elderly Widows

About 64% of elderly widows were in the young-old age group (60–69 years) followed by old-old age group (70–79 years) and oldest-old (80 and above years) (27 and 9%, respectively). Majority of the elderly widows were non-literates. Work status of the elderly widows indicates that only 30% of them were engaged in one or other form of economic activity at the time of survey. Among them, 30% of the elderly widows were engaged in agricultural work and 47% were non-agricultural labourers, 21% were self-employed and only 2% were salaried employees. More than half of the respondents (67%) were staying in nuclear families (Table 10.2).

Table 10.2 Distribution of elderly widows by selected socio-economic and demographic characteristics (as %)

Background characteristics	Per cent	Number
Age group		
Young-old (60–69 years)	64.3	193
Old-old (70–79 years)	27.0	81
Oldest-old (80 and above years)	8.7	26
Religion		
Hindu	80.3	241
Muslim	7.0	21
Sarna (cf. fn 1)	12.7	38
Caste		
Scheduled castes (SCs)	14.3	43
Scheduled tribes (STs)	40.0	120
Other backward castes (OBCs)	37.3	112
Others	8.3	25
Literacy		
Literate	9.7	29
Non-literate	90.3	271
Number of sons		
No son	21.3	64
One son	28.3	85
Two sons	27.0	81
Three or more sons	23.3	70
Number of daughters		
No daughter	21.7	65
One daughter	26.7	80
Two daughters	24.0	72
Three or more daughters	27.6	83
Current work status		
Working	29.7	87
Not working	70.3	213
Occupational status (among those working)		
Non-agricultural labour	47.1	41
Agricultural labour	29.9	26
Self-employed/Petty trade	20.7	18
Salaried	2.3	2
Total number of elderly widows	100.0	300

Living Arrangements of Elderly Widows

In India, family is the primary caregiver for elderly persons and acts as a principal source of support and security in old age. Living arrangements of the elderly, where the older people live and with whom they live, make a significant difference to a

Table 10.3 Per cent distribution of elderly widows according to living arrangements

Living arrangements	Percentage	Number
Living alone	13.0	39
Living with unmarried children	14.7	44
Living with married son	55.3	166
Living with married daughter	6.0	18
Living with other relatives	11.0	33
Number of elderly widows	100.0	300

variety of factors that may directly or indirectly contribute to the quality of their life (Prakash 1999) and also provides some indications of the level of actual support available to the elderly (Rajan and Kumar 2003). Due to the effect of modernization, westernization and urbanization in our society, the traditional bonds with the extended family members are now gradually breaking. The elderly have been the biggest sufferers of the changing values in the family system. In the Indian context, it is not only the number of children available, but their gender and marital status that also determine the co-residence pattern.

The distribution of elderly widows by their actual living arrangements is presented in Table 10.3. With regard to living arrangements of elderly widows, nearly 15% of the elderly widows lived with their unmarried children. More than half of the respondents stayed with their married sons compared to only 6% who lived with married daughters. This clearly reflects the typical characteristics of a patrilineal society, i.e. the cultural norm of elderly parents living with sons.

The elderly widows live with their married sons either together in a large household or adjacent to it. When this preferred arrangement is not possible, then the elderly widows prefer to co-live with others. It was found that 11% of elderly widows lived with their relatives, i.e. with brother, sister, grandchildren, etc. Thirteen per cent of elderly widows stayed independently without any direct support from family members (Table 10.3). On the other hand, living alone does not mean that potential support is altogether non-existent to the elderly widows. But it indicates that the amount of support is likely to be lower and the degree of loneliness and social isolation is also likely to be higher as compared to those who lived with children or other relatives (Panda 1997). It is also an indication that the traditional family system is gradually changing in India, even in remote rural areas.

The decision-maker about the place of residence for the elderly widows was also investigated. Among those who were currently living alone, 23% of elderly widows said that their children took the decision in this regard. However, for majority, it was their own decision to stay alone. Among those who were currently living with their married children or grandchildren, 33% said that their children took the decision, and in 9% of cases, the decision was made by their relatives (Table 10.4).

A question was asked to elderly widows—"*How do you feel about your current living condition?*" The responses were broadly categorized into three—*comfortable, satisfactory and uncomfortable*. Among those who are living alone, 46% felt that their current living condition was comfortable. Around 36% said that their current living condition was uncomfortable. More than half (63.6%) of the

Table 10.4 Per cent distribution of elderly widows by living arrangements and persons who took the decision

	Actual living arrangements			
Decision by	Living alone	With unmarried children	With married children or grandchildren	With other relatives
Self	69.2	81.8	58.1	14.3
Children	23.1	18.2	33.0	7.1
Other relatives	7.7	0.0	8.9	78.6
Number of elderly widows	39	44	203	14

respondents who were living with their unmarried children were comfortable, and 18% said that they were satisfied. However, equal percentage of elderly widows was uncomfortable with their current living conditions. Among elderly widows who were living with their married children or grandchildren, more than half of them were comfortable with their current living condition. Only 13% of elderly widows expressed that they were "uncomfortable" to stay with their married children or grandchildren.

While interviewing, an elderly widow said—"*After the death of my husband, I suffered a lot. After two years of my son's marriage, my daughter-in-law started quarrelling with me. She never gives me enough food*". Further she mentioned —"*For who I was doing all the hard work all these years, now treats me like this! My son and daughter-in-law will throw me out one day. The way they behave now, it is surely going to happen*".

A question was posed to elderly widows about their preference in living arrangements. Among the elderly widows who were currently living alone, 67% preferred to live alone and slightly less than one-fourth (23%) wanted to live with their married sons. Among the respondents who were currently residing with their

Table 10.5 Per cent distribution of elderly widows according to preferred living arrangements and actual living arrangements

		Preferred living arrangements			
Actual living arrangements	Living alone	With unmarried children	With married sons	With married daughters	With other relatives
Living alone	66.7	0.0	23.1	2.6	7.7
Living with unmarried children	9.1	81.8	9.1	0.0	0.0
Living with married son/s	7.8	0.6	91.0	0.6	0.0
Living with married daughter/s	0.0	0.0	5.6	94.4	0.0
Living with other relatives	15.2	0.0	9.1	0.0	75.8

married sons, majority preferred to live with them. However, 8% prefer to live alone. Among elderly widows who were currently living with their married daughters, only 6% preferred to live with their married sons. Among those who were currently living with other relatives, 15% preferred to live alone (Table 10.5).

Widowhood is a stage in life in which woman faces several difficulties after the death of her husband. In many instances, family members do not want to keep them. To understand the factors that influence the preference for living arrangements, the actual living arrangement and the stated preferences were compared. Around 47% of the respondents stated that there was nobody to take care of them. Thirty-five percent of elderly widows said that it was very difficult to manage everything at this stage of their life. Five percent stated that their daughters-in-law were not supporting them. As one old lady narrated her problem—"*My daughter-in-law always tries to find fault with me. Even if my son is eager to keep me with him, she is not allowing. What can I do*?" Around 13% also stated other reasons such as, "they don't have son/daughter", "it is not good to stay with married daughter", etc.

Economic Status and Dependency Level

The elderly expect economic, social and emotional support from family members. Their economic productivity and physical strength decline considerably with advancing age. Economic factors are major determinants of well-being. Economic necessities compel many elderly to continue to work with lower wages even in

Table 10.6 Per cent distribution of elderly widows according to work status and type of work

Work status	Per cent	Number
Ever worked		
Yes	66.0	198
No	34.0	102
Currently working		
Yes	29.0	87
No	71.0	213
Type of work[a]		
Non-agricultural labourer	47.1	41
Agricultural labourer	29.9	26
Self-employed/Petty trade	20.7	18
Salaried	2.3	2
Whether receiving cash as remuneration for work[a]		
Yes	88.5	77
No	11.5	10

Note [a]Among those who are currently working (N = 87)

Table 10.7 Per cent distribution of elderly widows by level of economic dependency

Economic dependency level	Per cent	Number
Fully dependent	40.3	121
Partially dependent	17.7	53
Not dependent	42.0	126
Number of elderly widows	100.0	300

unhealthy and hazardous environment, to take care of themselves and to support those directly depending on them (Sekher 2005). More than half (66%) of elderly widows were engaged in some income earning activity. However, only one-fourth (29%) of them were working at the time of survey (Table 10.6).

Further, elderly widows were asked—"*Why are you still working, even at this advanced age?*" Nearly three-fourths of them were forced to work to manage the household requirements. One-tenth among elderly widows continue to work to support their children. Fifteen percent of elderly widows were working to earn "*pocket money*", which they can spend for themselves. After husband's death, many elderly widows had to take the responsibility for earning and managing their households as well as maintaining children. Whatever little amount of money they earned was needed to support the family.

More than half (51.7%) of elderly widows were getting Rs. 101–500 in a month for their work, and about one-fifth (19.5%) were getting Rs. 501–1000. Around 13% were getting only Rs. 30–100 in a month for their work. Only 5% of elderly widows were getting more than Rs. 1000 in a month for their work. A sarpanch of one study village said—"*People prefer to keep old widows in their houses because they are ready to work even for no wages (just for food). Another reason for keeping elderly widows as maids is they are available full time in the house. Young maids prefer to take up work in many households to earn more*".

A question was asked to the elderly widows—"*Are you economically dependent on others?*" The responses were grouped into three broad categories—*fully dependent on others, partially dependent and not dependent.* Table 10.7 illustrates that around two-fifths (40.3%) of the elderly widows were fully dependent on others for economic support. Around 18% were partially dependent and 42% were not dependent on others (Table 10.7).

Multigenerational living arrangements have been an integral part of the rural social life in India. However, with changing attitudes and lifestyles, several young adults prefer to live in nuclear families even in rural areas. This shift in the attitude leaves many older women to fend for themselves, particularly after the death of their husbands. Sometimes family members economically support the elderly widows and sometimes they do not. Elderly are no longer considered as productive and useful. For many, an elderly widow is a burden on the family. So, it is important to know who provides economic support to the widows. Majority (45%) of the economically dependent elderly widows were supported by their sons. Thirty per cent of elderly widows stated that they were managing on their own. Interestingly, 12% of elderly widows were depending on their daughters for financial support. Few receive financial support from their siblings. However, around 36% were not satisfied with the economic support that they received from family members.

As one elderly widow said—"*Now-a-days, everyone wants to stay in a small family of husband and wife. Nobody wants their old parents to stay with them. Especially if one becomes widow, then nobody is there to take care of her*". Further she said—"*Duniya janta hai ki pati rahne se sabhi puchhte hai, baad mein koi nahi puchhta hai*" *(Everyone knows in this world, if husband is alive then everyone will ask about you, otherwise nobody even think of you)* (Refer case studies in Appendix: A).

According to the existing practice in the study area, land was usually inherited from father to son. According to customs, a widow possessed rights over the land of her deceased husband until her sons were mature enough to manage the family property. Eighty two percent of elderly widows do not have a house in their name, and half of the elderly widows do not own any land. When the question was asked about the ownership of the house in which they were currently living, 39% of the elderly widows said that the house was in the name of her deceased husband. Only 18% of elderly widows had the ownership of the house in which she was residing.

Physical Mobility and Functional Disability

Physical mobility is an important indicator of physical condition of health and is also an indication of the degree of dependence on others for movement and performing daily routine. The physical mobility was categorized into three—*physically immobile, confined to bed and confined to home*. It was found that majority of the elderly widows (69.3%) were physically mobile. However, 30% were confined to home and less than 1% was confined to bed. Functional disability was measured in terms of self-reported disability in routine daily activities which is captured by answering the following questions—"*Do you require any help or support for:*

Table 10.8 Percentage of elderly widows by activities of daily living (ADL)

Daily activities (ADL)	No need of any help	Need help	Can't do even with help
Getting out of bed	86.3	13.7	0.0
Bathing	90.7	9.3	0.0
Walking for some distance	74.0	26.0	0.0
Eating	97.3	2.7	0.0
Dressing	94.7	5.3	0.0
Toileting	83.3	16.7	0.0
Cooking	71.7	12.3	16.0

getting out of bed, bathing, walking for some distance, taking food, dressing, going to toilet, and cooking?" The answers were coded into three categories—*no need of any help, need help, and can't do even with help.*

Around one-fifth of elderly widows needed help for walking, 17% required help for going to toilet, 14% were dependent on others for getting out of bed and 16% could not cook food for self even with help (Table 10.8). This shows the extent of physical dependency of elderly widows for undertaking their day-to-day activities.

Psychological Health Status of Elderly Widows

To measure the level of depression or psychological condition of elderly widows, Geriatric Depression Scale (GDS) was used in this study. This scale is a self-reported 30-item inventory with simple "Yes/No" answers designed specifically to identify depression among elderly. The time frame for this scale is last one month. This scale has 84% sensitivity and 95% specificity at a cut-off point of 11. It also has high test–retest reliability and internal consistency (Yesavage et al. 1983). Twenty-four items have been identified from the 30-item GDS scale. In order to adapt the scale to the Indian situation, certain questions were excluded from the scale, as it may not be relevant in Indian context. Certain questions were modified as per the requirement and the scale was finalized. The total score is calculated by totalling one point counted for each depression answer. The score given for the construction of the present scale:

- Score of 0–8: no depression or normal
- Score 9–17: mild depression
- Score over 18: severe depression.

After checking the reliability with alpha values (0.6516), the composite index was computed.

Based on the GDS administered, it was found that around 47% of elderly widows had mild depression and 31% were severely depressed (Table 10.9). The government medical officer of a study village said—"*Loneliness and depression are*

Table 10.9 Depression among elderly widows

Depression	Percentage	Frequency
No depression	22.3	67
Mild depression	46.7	140
Severe depression	31.0	93
Cronbach's alpha value	0.6516	

common among the elderly widows. Depression leads to more serious health problems". According to the type of family of elderly widows, it was found that among those who were living alone, around 35% were severely depressed compared to those who were staying in joint families. "*It is the responsibility of family members to take care of the elderly and widows. But many never worried for the mental peace of their elderly parents*" *(Key informant interview with a medical officer).* Economically dependent elderly widows were more severely depressed than that of economically independent. (See key informant interviews in Appendix: B).

Living Arrangements and Level of Depression Among Elderly Widows

It is said that the presence of family members is very important as they provide emotional as well as instrumental support to elderly. An attempt was made to understand the level of depression according to the living arrangements of elderly widows.

Among the elderly widows who were living alone, around 36% were severely depressed (Table 10.10). Among elderly widows living with unmarried children, only 15% were not depressed. Among elderly widows who lived with their married children and other relatives, 46% was moderately depressed and 30% were severely depressed.

Elderly Widow's Relationship with Neighbours/Friends

An elderly widow (67 years) stated that—"*after the death of my husband, my brother's son has taken all responsibility. He takes care of me like his own mother. My brother and his family are doing everything for me but I do not have anything to*

Table 10.10 Level of depression among elderly widows by living arrangements

Living arrangements	Depression level			
	No depression	Mild depression	Severe depression	χ^2 value
Living alone	20.5	43.6	35.9	2.2*
Living with unmarried children	15.9	54.5	29.5	
Living with married children and other relatives	24.0	45.6	30.4	

Significance level given as $*P < 0.01$

give back to them in return". An attempt has been made to understand the elderly widow's relationship with their friends and neighbours. Majority of elderly widows stated that they preferred to share their problems with neighbours. However, 66% of the elderly widows do not have any close/intimate person with whom they can share their problems.

Elderly parents usually expect too much from their sons. Probing was done to know whether elderly widows were getting expected care and support from their son/s. Thirty eight percent of elderly widows were not getting the expected care and support from their sons. An opinion was sought from elderly widows about the benefit of having a son in their life. A question was asked to all elderly widows who do not have a son—"*Do you think your conditions would have been better, if you had a son?*" More than 80% of elderly widows who did not have a son thought that their condition would have been better if they had a son. However, around 16% felt that having a son would have made no difference at all in their living conditions. Which aspects of their life would have been better, if they had a son? About 48% of elderly widows replied that their sons would have taken care of them after husband's death. Around one-fourth of elderly widows said that if they had a son, their economic condition would have been better. Around 4% had revealed that there would not have been any shelter problem for them.

Elderly Widow's Participation in Social and Religious Activities

Information was gathered about the participation in social and religious activities of elderly widows. Half of the respondents usually participate in family-get-together. Half of them do participate in social gatherings such as marriage, funeral or any other functions. Some of the elderly widows revealed that some people never invite them to social or religious functions.

While interviewing, one elderly widow said—"*I attend marriages in our village. But when the new bride comes to the house, I am not allowed to go there and see her*". She continued "even *if the new bride comes to our house, I cannot perform any rituals*". It is evident that more than half (61%) of elderly widows do not visit or participate in any functions outside village. Around 27% of elderly widows were not allowed to go out of the village to participate in any functions. Forty-one per cent of elderly widows were allowed to go out of the village, only when accompanied by someone.

Decision-Making by Elderly Widows

One way to understand the status of elderly widows is to know their decision-making role on various issues. An attempt was made to understand the status of elderly widows by their ability to take decisions. Elderly widows were asked about the type of decisions in which they were usually consulted by the family members. It was found that around 48% of elderly widows took their own decisions to visit places outside the village. Forty-two percent of elderly widows revealed that their sons or daughters-in-law usually took decisions on their behalf regarding their visits outside the village. Forty-two percent of elderly widows said that they were not consulted while taking decisions. One-fourth (23.7%) of them said that during social occasions, their family members usually consult them. However, only 12% of elderly widows were consulted on economic matters.

To know the changing status of elderly widows before and after their widowhood, further probing was done to know whether their suggestions were taken seriously on family matters. More than half of elderly widows reported that their opinions were taken seriously on family matters when their husband's were alive. But after husband's death, majority were not consulted on any family matters. This shows the changing status of women after becoming widow and reduced decision-making power within their family.

To examine the status within their own households, a question was asked—"*In your opinion, what is your status within your family?*" The responses were categorized into three—*low, medium and good.* Majority felt that their status is low in the village, and only one out of seven felt that widows have good status in villages. When asked about the factors responsible for the low status of elderly widows, majority said that economic constraints were the main factors for the poor status of elderly widows. However, one-fourth believed that social factors were responsible for the low status of widows. The state of widowhood is exacerbated by conditions of poverty (United Nations 2001). Around 81% of elderly widows had faced many problems immediately after becoming widow. More than half of the elderly widows had faced economic problems.

In many rural communities, women were blamed for the death of their husbands. A 65-year-old tribal widow revealed—"*after my husband's death, people said that I am a witch and I only killed my husband. My neighbour one day warned me—if you want to be alive, then do not show your face to anyone and stay inside your home*". Widowhood is still considered as a curse for women in villages. They were exploited and humiliated by villagers for no reasons.

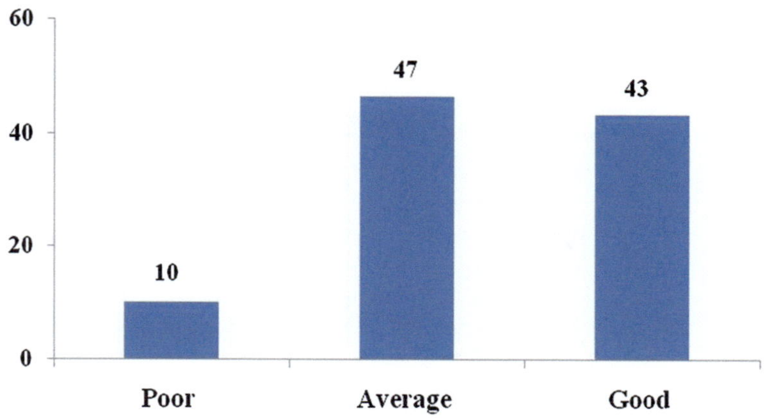

Fig. 10.1 Self-adjustment of elderly widows (in percentages)

Self-Adjustment of Elderly Widows

A self-adjustment inventory has been used in this study to understand how elderly widows were adjusting with their life problems. To assess the self-adjustment, a set of questions were asked. The statements were framed in such a way that it could be answered either as "*yes*" or "*no*". The questions such as—"*Do you think that you are much better than many other women of your age? Do you have any hobby to engage your time?*" were asked to understand the adjustment among the elderly widows. The questions in the self-adjustment inventory were modified to suit Indian rural situation. One point is given to each answer which shows the adjustment among the elderly widows. After checking for the reliability test, the alpha value comes to 0.5368. The scale of self-adjustment has been categorized as 0–4 points: "*Poor*", 5–8 points: "*Average*" and 9 and above points: "*Good*". Forty-three percent of the elderly widows were found to be "good" in self-adjustment, whereas 10% were "poor" in self-adjustment of their life problems (Fig. 10.1).

Social Vulnerability Among Elderly Widows

An attempt was made to understand whether the elderly widows were getting due respect from the younger generations. Thirty-one percent of elderly widows felt that younger generation was not showing the respect they really deserved. This opinion is mainly based on the respondent's comparison of the situation at two points in their lifetime—when they were young and when they were old. They naturally tend to compare the extent of respect they have shown to their elders and the type of respect that they now receive from youngsters. Further, a question was asked—"*Do*

Table 10.11 Social adjustment inventory among elderly widows by selected background characteristics

Selected characteristics	Social adjustment				
	Low	Medium	High	Number	χ^2value
Age group					
Young-old (60–69 years)	10.4	38.3	51.3	193	3.8
Old-old (70–79 years)	3.7	40.7	55.6	81	
Oldest-old (80+ years)	7.7	46.2	46.2	26	
Work status					0.6
Currently working	6.9	37.9	55.2	87	
Currently not working	8.9	40.4	50.7	213	
Type of family					3.9
Single	15.4	33.3	51.3	39	
Nuclear family	8.0	41.0	51.0	200	
Joint family	4.9	39.3	55.7	61	
Economic dependency					10.9
Economically dependent	4.0	43.7	52.3	174	
Economically not	14.3	34.1	51.6	126	
Health status					7.4*
Very good/good	6.7	35.0	58.3	60	
Fair	8.1	33.9	58.1	124	
Poor/very poor	9.5	48.3	42.2	116	
Emotional deprivation					34.3***
Normal	2.3	29.1	68.6	86	
Low deprivation	1.3	37.3	61.3	75	
High deprivation	15.8	47.5	36.7	139	
Total	8.3	39.7	52.0	300	

Significance level given as $^*P < 0.1$, $^{***}P < 0.01$

you think, widows are not properly treated in your village?" More than half of the elderly widows consider that widows are properly treated by the villagers. Most of the elderly widows (60.7%) blame their own fate for the present pathetic condition. Around two-fifths blamed their own family members for the problems.

Social Adjustment Inventory of Elderly Widows

Social adjustment refers to the subjective satisfaction with relationships or performance in social roles (Shapiro and Keyes 2008). In spite of social exclusion, widows try to adjust with the society in which they live. To understand the social adjustment among the elderly widows, social adjustment inventory was employed. A set of questions consisting of 16 items were canvassed and one point was given to each answer, which shows the social adjustment among them. Out of 16 items, 2 items have been dropped as these two were not showing significant correlation with the other variables. So, 14 items were used for the computation of composite index of social adjustment. The reliability with alpha values comes to 0.6315. The score has been categorized into three—"*low*" (0–4 score), "*medium*" (5–9 score) and "*high*" (10 and above score) levels of social adjustment. Fifty-twopercent of elderly widows were found to have good social adjustment, according to this study. No significant correlation was observed between the work status and social adjustment of elderly widows (Table 10.11). Fifteen percent of elderly widows living alone showed low social adjustment as compared to 4% who lived in joint families. Fourteen per cent of economically independent elderly widows were at low social adjustment. Among those who had good or very good health conditions, more than half of them indicated high level of social adjustment. Emotional deprivation among elderly widows has shown significant co-relation with their social adjustment.

Conclusions and Recommendations

The study revealed that elderly widows face discrimination at the household and at the community levels which varies from economic, physical to emotional needs. Being aged as well as widowed, they do not have intimate persons with whom they can share their problems. Many elderly widows experienced loneliness after the death of their husbands. Along with poverty and economic constraints, mistreatment and neglect by the family members make them further vulnerable. Most of them silently cope with this situation.

The traditional bonds which existed in the extended family were now gradually breaking. The elderly have been the biggest sufferers of the changing values in the family system in India. Caring for the elderly was not considered a serious issue in India till recently, as they were provided with social protection, in the joint family living arrangements (Rajan et al. 1999). But with the breaking down of joint families, many elderly widows were left to fend themselves.

In the Indian context, it is not only the number of children available but their gender and marital status that also determine the co-residence pattern. It is also an indication that the traditional family system is gradually changing, even in remote rural areas. Around one-fourth of elderly widows lived alone when they did not have a daughter. This is one of the indications of the importance of sons in determining the living arrangements of elderly. Other studies also supported the view that the number of children positively related to the probability of living with them (Rajan and Kumar 2003).

The main dimensions which impact the well-being of elderly widows are economic, health and social aspects. There are other aspects such as education, status and respect, community activities, and societal attitudes, which also indirectly influence their quality of life in old age. Welfare measures for the elderly, especially for widows, need to be developed. Considering the low levels of literacy among them, their engagement in the unorganized sector, poor financial security in terms of possession of assets and property, and their overall low standard of living, appropriate welfare programmes need to be designed and implemented.

Loneliness many times leads to depression and isolation of widowed women. So, there should be some support system, especially for those who are suffering from physical or mental illness. Sometimes widows are discouraged from taking part in auspicious ceremonies and social activities. Even today, some widows in rural areas are accused of being responsible for the death of their husbands. Based on the realization that widowhood is generally stressful for women, it is recommended that massive and intensive campaign should be mounted to sensitize people on the need to stop all forms of abuse and discrimination associated with widowhood. Widows need to be provided with functional education capable of emancipating them from the stress and tensions of widowhood. Policies and programmes should ensure that older persons have a reasonable and adequate living environment. For this, incentives should be given to facilitate appropriate housing schemes for elderly widows who live alone, either by choice or by circumstances.

Appendix A

Case Study 1

Mrs. K is 83 years and has two sons and four daughters. She got married in her childhood (she does not even know the exact age of her marriage). When she became 15 years, she came to her in-law's house. She and her husband were working as agricultural labourers. They were cultivating vegetables and selling in nearby market. She also got 25 katta of land from her father. With this land and hard work, they managed to give education to their children and arranged their marriages. For her daughter's marriage, they have to sell 10 katta of land to pay dowry. Her elder son was a vegetable vendor but was not earning enough money to run a house. One day, he committed suicide. After that, daughter-in-law took all the land and continues to stay in the house.

Mrs. K underwent very difficult times, neglected by family members. Any how she got a BPL card with the help of other villagers in her own name. With this BPL card, she is eligible to get 21 kg of rice, 13 kg of wheat and 3 litter of kerosene every month. She said—"*although I am not getting all these items every month, I get rice every month*". She keeps 5 kg of rice for herself and sells the remaining 16 kg to earn some money. With this money, she is managing her other expenses.

She is not receiving widow pension or old-age pension. Even she never applied. She said that it is very difficult to get any money from government as lot of paper work is needed—"*I cannot do this paper work all alone*". She had some money saved in her name (Rs. 12,000) which she got from her land. She said—"*I cannot use this money because if I use it, then any emergency happens, whom I will ask?*" She cannot expect anything from her son and daughters-in-law. Further she mentioned—"*My daughter-in-law always quarrel with me. My son and daughter-in-law are not taking care of me. My neighbours help me*". In her own words—"*Isi din ko dekhane ke liye itana kast karke bete to aadmee banaye, pr mujhe kya pata that ki beta aisa ho jayega*" *(I have taken all pains to bring up my son. But for whom I was doing all the hard work all these years, now treats me like this!).* One day her son and daughter-in-law started abusing her and started beating her. When her neighbours came to stop them, her son said to them—"*this is our family matter, no need to interfere in our matters*". Next day, one of her neighbours advised her to complain to the police and then she lodged a complaint against her own son and daughter-in-law. Police came and took her son. Her daughter-in-law started threatening the old widow. Finally, she went to police station and requested the officer to leave her son. After sometime, her son started beating her again. "*I cannot go to the police station again. Even my neighbours will not help me*".

She said with tears in her eyes—"*I will die, but I will not leave this house*". She is even worried that her son will question her for talking to "outsiders and strangers". "*This house is like a jail. I have no freedom to talk, move and meet people. At least in jail, prisoners get enough food to eat*". Her words reflect the pathetic condition of elderly widows.

Case Study 2

Mrs. N is 70 years and she has two daughters and two sons. Both the sons and daughters got married and settled. Her sons are working in Tata Steel Company and staying in the city. Her husband was an agricultural labourer. They had a small piece of land and one small shop. When her husband was alive, they were staying in the village. Most unexpectedly, her husband passed away, leaving her alone.

After her husband's death, she was staying alone in the same village. After her husband's death, her elder son requested her to stay with him. One year she stayed with the elder son and six months with the younger son. She said—"*my sons are good, but daughters-in-law are not that good. In the beginning my daughters-in-law were taking care of me, but gradually they started quarrelling with me on small things. Then, I told my son that I want to go back to the village. In the beginning, he was not ready to send me back to the village. But somehow I convinced him telling that someone should be there to look after the house and land*".

She said—"*Now-a-days, everyone wants to stay in a small family of husband, wife and children. Nobody wants their old parents to stay with them. Especially if one becomes widow, then nobody is there to take care of her*". Sons come to see her every month and give her required money. One of her grandchildren was staying with her since his childhood. But now he has grown up and got married. Her grandson and his family are living very close to her house. "*Jise chhote se bara kiye wo bhi dekh-bhal nahi karta hai, bahu ke karan muje yanha rahana parta hai, kya kare jeene ke liya to rahana parega na*" *(I have taken care of him for many years, but he is also not taking care of me. See, I have to stay here alone).*

Her nephew was taking care of her land but he never gives money or rice from the land to this old lady. She shares all her problems with one of her neighbours. She cannot walk properly, so her neighbour visits her every evening. For vegetables and other requirements, villagers help her. She said—"*now I do not want to live more. It is better to die rather than staying like this*". She continued—"*Duniya janta hai ki pati rahne se sabhi puchhte hai baad mein koi nahi puchhta hai*", *(Everyone knows, if husband is alive then everyone will ask about you, otherwise nobody will even think of you).* "*Being a widow, I do not have any problem as such. But having two sons and daughters-in-law, still I am staying alone, this makes me feel more depressed*". Sometimes, even neighbours blame her. "*Why are you staying here alone? You should go and stay with your sons. You have every right to demand help and protection from your adult sons*". The loneliness and helplessness of this old lady is very much evident in her words and action.

Appendix: B: Key Informant Interviews

Sarpanch (Head) of the Village

Mr. W is sarpanch of Village X for the last 13 years. Regarding the status of elders in the village, sarpanch said—"*during my young age, we used to give lot of respect to elders, but the new generation does not pay any attention towards elders*". Now-a-days the young daughters-in-law are not taking care of their mothers-in-law. The situation is more difficult, in case the woman is a widow. In many families, sons and daughters-in-law never want to support their elderly mothers.

Further, he mentioned—"In this village, mostly poor widows are working as "*maids*" in rich households. People want to keep them as they will not have much responsibility for their children, and they are *ready to work at low salary*. Another reason of keeping elderly widows as maids is they will work full time without asking extra money. Since they are old, they usually do not go away from the village".

He further mentioned that earlier widows in some of these villages are viewed as "*daayan*" (witch), but now the attitudes of people are changing and they are not blaming widows for the death of their husbands. But still "*daayan pratha*" exists in the community and this is more common among the lower castes. Lower caste widows are having more freedom than those from upper castes. Higher castes are more rigid on their cultural practices and have more restrictions for widows. However, widows from the lower castes can re-marry. Even then there is not a single case of re-marriage in the village or neighbouring villages. In case any young widows re-marry, then she will be thrown out from the village and community. Sarpanch admitted that many deserving elderly widows are not getting pension. According to him, he went many times to the block level offices enquiring about widow/old-age pension. "*How can they be so insensitive to poor old women?*" he asks. According to him, though there is old-age home in the district, many elderly widows are not interested to go and stay there, even if it is free of cost. "*Old people are so attached to their house and village; they do not want to stay anywhere else. They want to die in the same place where they spent all their life*". Even well-settled children living in towns want to take elderly mothers to their place, but women are least interested to move out of the village.

Doctor in a Government Hospital in the Study Area

According to the Medical Officer of a study village, most of the elderly patients complain about general health problems arise during old age such as joint pain and weakness. Some elderly are suffering from asthma, hypertension and diabetes. Family members do not understand the health requirements of elderly members which demand regular check-ups. Sometimes family members bring elderly at the last stage when the health condition is very serious. For those elderly or widows who are staying alone, nobody is worried. Even their children staying in the same village never bothered to find out their health problems.

Old age, widowhood and staying alone add mental tension leading to depression among old widows. Loneliness and depression are common among the elderly widows in the locality. According to the doctor, the relatives are not ready to spend money for treatment of aged. "*When elderly comes alone, we request them to bring some family members along with them*". Many elderly widows are anaemic as they are poor and cannot afford nutritious food. "*Even the elderly living with their family never come for the routine check-up. Then what will be the situation of poor elderly widows and of those who are living alone*".

Family may not be always considerate in dispensing care to the elderly and this is more so when he/she becomes liability, physically and economically. Sometimes, elderly widows come to the hospital with minor problems. "*When I discuss with them, they start crying saying that nobody takes care of them*". Most problems are due to loneliness and fear. There are no psychiatrists in the block. But what hurts the elderly most is the way their family or children behaves with them. They realize that they are the most ignored and unwanted persons in the family. Depression gradually leads to more serious health problems. He mentioned—"*It is the responsibility of family members to take care of the elderly or widows. But many never worried about their own elderly parents*". There is a need for opening counselling centres for the aged and widows. Through proper counselling, many psychological and health problems can be minimized to a great extent.

According to the doctor, "*Elderly widows in the villages are the most neglected one. Nobody really care for them". Once a young man came to the hospital with his 80 years old mother who had severe health problems. Out of desperation, he told me—"Doctor, I am spending lot of money for my mother's treatment. I think, it is good for her and good for me, if she leaves this world as early as possible*".

References

Alam, M. (2006). *Ageing in India: Socio-economic and health dimensions*. New Delhi: Academic Foundation.

Census of India. (2011). *Social and cultural tables India*. Registrar General and Census Commissioner of India.

Chen, M., & Dreze, J. (1992). Widows and health in rural India. *Economic and Political Weekly, 28*(43, 44), WS 81–92.

Gupta.I., & Sankar, D. (2002). Health of the elderly in India: A multivariate analysis. *Discussion Paper Series No.—45*. New Delhi: Institute of Economic Growth.

Holmes, J. H., & Rahe, R. H. (1967). The social re-adjustments scale. *Journal of Psychosomatic Research, 11,* 213–228.

Kumari, S., & Sekher, T.V. (2010). Elderly in Punjab: Economic dependency, health conditions and living arrangements. *Man & Development, XXXII* (1), 77–88.

Kumari, S., & Sekher, T.V. (2012). Status of elderly widows in india. *BOLD*, *22*(2), 6–13.

Momtaz, Yadollah Abolfathi, Hamid, Tengku Aizan, Yahaya, Nurizan, & Ibrahim, Rahimah. (2009). Widowhood and psychological well-being among older Malaysians mediating effect of social network. *Indian Journal of Social Work, 70*(3), 375–390.

Panda, P. K. (1997). *Living arrangements of the elderly in rural Orissa, Working Paper.* Thiruvananthapuram: Center for Development Studies.

Prakash, I. J. (1999). Living arrangements of the elderly in India: Implications for care giving. *Ageing and Society, 9*(3–4), 1–11.

Rajan, S., Irudaya, U. S. Mishra, & Sharma, P. S. (1999). *India's elderly: Burden or challenge?*. New Delhi: Sage Publications.

Rajan, S. Irudaya, & Kumar, Sanjay. (2003). Living arrangements among Indian elderly: New evidence from national family health survey. *Economic and Political Weekly, 38*(1), 75–80.

Sebastian, D., & Sekher, T. V. (2011). Extent and nature of elder abuse in Indian families: A study in Kerala. *Research & Development Journal, 17*(3), 20–28.

Sebastian, D. (2011). Abuse and neglect of the elderly in Indian families. *BOLD, 21*(4), 5–12.

Sekher, T. V. (2005). Socio-economic dimensions of old age security in India: With special reference to Karnataka. *Journal of Social and Economic Development, 7* (1), 12–28.

Shapiro, Adam, & Lee Corey, Keyes M. (2008). Marital status and social well-being: Are the married always better off? *Social Indicators Research, 88,* 329–346.

Troisi, J. (2000). *Tribal religion*. New Delhi: Manohar Publishers.

UNFPA. (1998). *The state of world population-1998*. New York: United Nations Population Fund.

United Nations. (2001). *Widowhood: Invisible women, secluded or excluded, woman 2000 and beyond*. United Nations, New York: Division for the Advancement of Women.

United Nations. (2002). *World population ageing 1950–2050*. New York: Population division, Department of Economic and Social Affairs.

Yesavage, J.A. Brink, T.L Rose, T.L. Lum O, Huang V, Adey M.B, & Leirer, V.O. (1983). Development and validation of a geriatric depression screening scale: A preliminary report. *Journal of Psychiatric Research, 17*, 37–49.

Chapter 11
Elderly Women: Prevailing Paradigm of Caring Scenario in the Backdrop of Feminisation of Ageing

S. Jayashree

> *Sir James Sterling Ross commented,*
> *You do not heal old age. You protect it, you promote it, and you extend it.*

Abstract Ageing is a demographic reality which poses many challenges for the aged as well as their caregivers. Though ageing is a natural and universal phenomenon, it impinges differently for men and women due to biological and cultural factors. One of the glaring issues of population ageing is feminization of ageing and longer survival of women poses lot of challenges in terms of care, service, finance, health, widowhood and morbidity. The study on care among old-old widows was conducted in Dharwad town. The research identified three areas where the problem is conspicuous—support system, health care and living arrangement. Since women live longer than men and the extended period of life is spent more or less in morbidity and neglect, it is suggested that the amount of old-age pension should be higher for women than for men.

Keywords Old–old · Widow · Care · Social support · Living arrangements · Social network

Introduction

Ageing is a demographic reality which poses many challenges for the aged as well as their caregivers. Though ageing is a natural and universal phenomenon, it impinges differently for men and women due to biological and cultural factors. With the reduction of fertility, mortality and enhanced longevity, India is experiencing rapid shifts in the age structure of its population. The ramifications of population ageing affect various social, economic, health issues of elderly on a day-to-day basis. Over the last two decades, there has been a tremendous growth of

S. Jayashree (✉)
Department of Sociology, Karnatak University, Dharwad 580003, Karnataka, India
e-mail: jaishiva678@gmail.com

S. Irudaya Rajan and G. Balagopal (eds.), *Elderly Care in India*,
DOI 10.1007/978-981-10-3439-8_11

elderly population in terms of number and proportion. The proportion of population aged 60 years and above was 8.6% in 2011. It is projected to increase to 12.4% by 2026 (Office of the Registrar General and Census Commissioner, 2006). One of the glaring issues of population ageing is feminization of ageing, wherein older women outnumber older men. Longer survival of women poses lot of challenges in terms of care, service, finance, health, widowhood and morbidity. Besides, these various "izations" and changing trends directly affect family values, living arrangements and lifestyles of people, which in turn impact elderly women's living arrangements, health care, financial assistance, social and emotional support. Thus, multiple marginalizations are evident for women who survive for longer years.

Review of Literature

There are host of studies pertaining to the pathological aspects of elderly in general and elderly women in particular. In case of elderly women, widowhood and menopausal issues were studied extensively these studies treated all widows as a homogeneous group. Despite their multiple marginalizations (being woman, being old, widowed and dependent), studies are deficient, in case of old-old elderly widows, their caring scenario and woes.

In India, family is the core-caregiving institution which takes care of elderly. Hence, unlike Western society, institutionalization is not in vogue in case of the elderly in India. Thus, care provided by adult children can prevent or postpone institutionalization (Dunlop 1980). Jamuna (1999a) highlighted the factors of quality of care of the elderly. Elder care becomes arduous and strenuous, if care is extended for a long period. She stressed appropriate incentives with an aim to maintain family care. Informal caregivers are by no means a homogeneous group and different types of caregivers have different needs. The caregiving demands often impinge upon the work, lives and other personal commitments of caregivers who are most typically middle-aged men and women.

Numerous studies have shown that adult children, particularly daughters, often provide long-term health care for their elderly parents. (Shanas 1979a, 1979b). Haynes (1976) notes that family influence is considerable, with "supportive" families being associated with greater compliance. Noncompliance and medication errors are associated with old age, perhaps because the elderly often experience problems of forgetfulness or self-neglect and tend to be treated with multiple medications for more than one condition. Thus, the provision of social support is especially relevant for this group. Lopata (1979) has defined a support system as a mutual, but not necessarily symmetrical, exchange of resources between individuals and between individual and groups in society. Exchanges take place in economic, service, social and emotional areas. The individual is the organizing agent who selects and negotiates a network of relations that, when activated, becomes a personal support system.

Informal support system refers to those among primary group members (family, friends and neighbours) as opposed to the formal support system (groups, organizations and agencies of a bureaucratic nature). Informal support system provides for both expressive and instrumental needs, whereas the formal support system provides mostly for instrumental needs. DiMatteo et al. (1981) suggested that social support during serious illness is most effective, when it is produced by a combination of primary groups (friends and family) and secondary groups (peer groups and helping professionals). Jayashree (2013) revealed that during old age, men are more dependent on women for their ADL and IADL activities. Though women suffer from structural disadvantages such as gender discrimination, negligence or widowhood, they were not dependent for their ADL and IADL activities. Thus, caring scenario differs with regard to gender.

Methodology

The study was conducted in Dharwad town. According to Census of India, the population of elderly in Karnataka increased from 7.3% in 2001 to 9.5% in 2011. Out of the total elderly population, 47.6% were elderly males and 52.3% were elderly females which clearly demonstrated feminization of ageing (Census of India 2011).

Dharwad town has 22 wards, from which 25% (6 wards) were taken for the study through lottery method. The study mainly focused on elderly (old–old) widows. At first, we encountered some problems in identifying them; finally with the help of voters list of these wards, we could obtain the age and marital status of the respondents. We got a total of 97 elderly widows who were between the ages of 70–80 years from six wards which constituted our sample. However, we could gather information from 88 elderly women, which constituted the final sample. The study was conducted during 2011–12.

Semi-structured interview schedule was used to elicit information from the respondents. Interview schedule mainly consisted of close-ended questions; however, open-ended questions were also used to probe further. Some of the unique cases have also been noted down during field work.

The objectives of the study are as follows:

1. To study socio-economic profile of the respondents.
2. To examine the prevalence of caring scenario of elderly women.
3. To understand the living arrangements and support system.
4. To ascertain the needs of elderly women and make suggestions.

Socio-Economic Profile of the Respondents

The study concentrated on widows who were between the ages of 70–80 years and 72.72% of them belonged to age group of 70–75 years (Table 11.1). Majority of them were illiterate and engaged in household activities. Majority of the

Table 11.1 Socio-economic profile of respondents in Dharwad

01	**Age of the Respondents**	**Frequency**	**Percentage**
	70–75	64	72.72
	76–80	24	27.28
02	**Education**		
	Illiterate	61	69.30
	Primary	14	15.90
	Secondary	02	2.27
	P.U.C.	06	6.80
	Degree	02	2.27
	Post graduation	03	3.40
03	**Occupation**		
	Household	78	88.60
	Retired from Govt. service	06	6.80
	Business/self employed	02	2.30
	Retired from private service	02	2.30
04	**Income**		
	Below 1000/-	52	59.09
	1001–3000/-	21	23.86
	3001–6000/-	11	12.05
	6001 and above	04	4.54
05	**Household**		
	Joint	08	9.10
	Nuclear	80	90.90
06	**Living Arrangements**		
	Living alone	03	3.40
	Living with son	52	59.10
	Living with daughter	13	14.77
	Living with unmarried son	03	3.40
	Living with relatives	14	15.90
	Living with friends	01	1.13
	Others (Living by rotation)	05	5.70
07	**Category**		
	General (Upper caste)	27	30.68
	Other backward class	29	32.95
	Schedule caste/Schedule tribe	18	20.45
	Others	14	15.90
08	**No. of Children**		
	01–03	26	29.54
	04–06	58	65.90
	07–10	04	4.54
09	**Duration of widowhood status**		
	Below 10 years	16	18.20
	11–20 years	68	77.30
	20 years and more	04	4.50

respondents belonged to households in which income was below Rs. 1000/-. Majority of the respondents lived with their son. Living with daughter and living by rotation were also found in the sample. A large proportion of the respondents (65.90%) had 4–6 children and 77.30% of the respondent's widowhood status ranged from 11 to 20 years.

Caring Scenario in the Family

Family care for older women is a vital concern. The traditional joint family system is evading mainly due to migration of young adults to cities leaving behind their parents/grandparents. Migration along with various types of "izations" has far-reaching implications on life of the elderly in general and elderly women in particular. Besides, these changes in size and composition of the household, dependency, loneliness, lack of social support and widowhood are added to the woes of the elderly women.

Although large proportion of the population lives in the rural areas, there is a trend of increased mobility of young adults from the rural areas to urban areas for making a living. According to the Census of India, 2011, 71% elderly resided in rural and 29% in urban areas (Census of India 2011). This trend of migration has not only certain economic benefits, but also some drawbacks such as contributing to the nuclearization of families and leaving behind ailing elderly parents/grandparents. Now, families are being pressured by too many demands; hence, elder care becomes of secondary importance in the family.

Enhanced longevity adds to the sufferings of elderly women. Women are the traditional "core caregivers" to family members. However, due to feminization of ageing, women live longer as compared to men. Women survive and serve the members of the family, when it comes to her care she has to depend on others. Thus, due to feminization of ageing women face a lot of hurdles and obstacles to lead a decent life at fag end of their life.

Those who are frail and in need of support, receive less care, and those who are more active and physically strong are less available to provide support. Caring has different dimensions and connotations. When it comes to older people, care usually means providing physical, social, economic, emotional and instrumental support on a continuous and long-term basis (Jaiprakash 2003). Caregiving is a family affair. This requires understanding and preparation from both caregivers and care receivers. Both need to be educated.

Even policy makers also value these familial systems of care for elders and prefer to maintain them rather than introducing other, potentially expensive, government programmes (Knodel, Chayovan, and Siriboon 1992). It is a culturally prescribed norm and primary duty of members of one's household to provide familial support for older adults.

In the backdrop of these developments, the present study collected information on caring scenario of the respondents on the basis of main family members who were taking care of elderly women in various activities. Majority of the respondents did not depend on others for their daily activities. Finance- and expenditure-related activities were taken care of by their sons. Respondents shared their emotional and expressive issues with their daughters. (Table 11.2) Those who do not have daughters were close to their daughters-in-law.

Only 17.4% of the respondents had very good sustaining relationship with family members and these respondents were financially sound, as they received pension/rent and had other sources of income (Table 11.3). Some of the respondents had property litigation, quarrel and conflict with family members, and they were fed up with their own kith and kin. Majority of the respondents were not ready to answer the questions about whom they shared their joys with. They kept mum. One of the respondents said: *"what joys of life I have now, my life is almost come to an end; you have come to listen to my despair and woes that is the only joy for me*".

Respondents were asked about their negative feeling/reasons for sadness. Respondents reported that financial crunch (93.18%) and dependency on family members (90.9%) were the twin issues about which they felt very sad (Table 11.4). Some of the respondents vociferously pointed out that, *if we have money we will be revered. Today money is everything, we are like empty pots. Who cares for us?* A large proportion (88.63%) of them were sad because of the careless attitude of the family members. It is very interesting to know that they were not as bothered about widowhood and advancing age. They have achieved "***role stability***" with regard to their widowhood. Thus, study indicates that respondents did not face any difficulty or hassles with regard to widowhood and they have successfully imbibed the role of widowhood. Because, respondents have lost their life partners when they were in the age range of 50–60 years, they had already spent 15–20 years without their husband. They learnt to manage their routine life and they internalized their widowhood. One of the respondents pointed out that, ***for woman widowhood is a matter of outward look and inward security***. Earlier there were many restrictions imposed on widows. She said *I put bindi, wear colourful sari with flower design which were banned earlier. Except for certain religious functions, rest of the functions I actively participate: like me there are many widows who are also attending functions. Death is inevitable. People have passed away are lucky we have to lead the life. For woman,* **role change is inevitable**. *We have left our parents who have reared, cared and nurtured us. Due to this, internalization of roles is not at all difficult, for women.* She blamed society which has seen widows as a different segment and ill-treats widows as second-grade citizens. Thus, the study found that dependency and financial crunch were the main reasons for despair for them rather than cultural and social constraints of widowhood.

Majority of the respondents (55.68%) opined that they do not have any active role in the family; meagre numbers of respondents helped their grandchildren (Table 11.5). Nearly, one-fifth (22.72%) of the respondents engaged in petty household work. Some respondents provided financial assistance to their family.

An overwhelming majority of the respondents revealed that family members grumble while giving care (Table 11.6). Indirect talk, neglect, not giving food on time, not preparing diet food, not bringing medicine, unnecessarily beating

Table 11.2 Primary caregivers of the elderly respondents

Persons taking care	Daily activities		Sharing of joys		Sharing of pains		Taking to hospital		Financial support		Emotional closeness		During ill health		Outings	
	F	%	F	%	F	%	F	%	F	%	F	%	F	%	F	%
Son	20	22.72	22	25.00	18	20.45	35	39.77	36	40.90	20	22.72	39	44.31	08	9.09
Daughter	10	11.36	10	11.36	35	39.77	08	9.09	21	23.86	42	47.72	22	25.00	19	21.59
Daughter-in-law	10	11.36	16	18.18	09	10.22	14	15.90	05	5.68	05	5.68	06	6.81	02	2.27
Unmarried Son	01	1.13	02	2.27	02	2.27	01	1.13	02	2.27	02	2.27	01	1.13	–	–
Relatives	08	9.09	06	6.81	08	9.09	05	5.68	06	6.81	04	4.54	07	7.95	03	3.40
Friends	01	1.13	04	4.54	05	5.68	01	1.13	–	–	01	1.13	01	1.13	–	–
Others/Self servants	38	43.18	–		09	10.22	10	11.36	18	20.45	–	–	–	–	–	–
No answer	–		28	31.81	03	3.40	14	15.90	–	–	14	15.90	12	13.63	56	63.63

Note F—Frequency; %—Percentage

Table 11.3 Overall relationship with family members

S. No.	Responses	Frequency	Percentage
01	Very good	15	17.04
02	Good	18	20.44
03	Neither good nor bad	46	52.28
04	Bad	04	4.54
05	Very bad	05	5.70
	Total	88	100.00

Table 11.4 Occasion of extreme sadness of respondents (N = 88)

S. No.	Occasions	Frequency	Percentage
01	During ill health	78	88.63
02	Financial crunch	82	93.18
03	Because of age	16	18.18
04	Dependency	80	90.90
05	Widowhood	41	46.59
06	Neglected by family members	78	88.63

Table 11.5 Active roles performed in the family

S. No.	Roles performed in the family	Frequency	Percentage
01	Engage in petty work	20	22.72
02	Helping grand children	06	6.80
03	Financial assistance	06	6.80
04	Emotional support	07	8.00
05	No role to play	49	55.68
	Total	88	100.00

Table 11.6 Grumbling of family members towards care

S. No.	Responses	Frequency	Percentage
01	Always	74	84.10
02	Sometimes	06	6.81
03	Never	08	9.09
	Total	88	100.00

grandchildren, rift and quarrelling were some of the manifestations of disrespect and grumbling nature. One retired teacher said *"In my home, if child refuses to eat something, my daughter-in-law grumbles, "what all I have to do for you. Whatever I prepare, you should eat and keep quite". It is not for her child but it is indirectly pointing to me".*

It is stunning to note that 75% of the respondents felt that old-age home would have been better than home (Table 11.7). One of the respondents said, *"After I lost*

Table 11.7 Opinion about old-age home

S. No.	Responses	Frequency	Percentage
01	Yes	66	75.00
02	No	15	17.04
03	NA	07	7.95
	Total	88	100.00

my husband I sacrificed a lot for my children by neglecting my basic needs and comforts. Now, my sacrifice has no value at all. Children are engrossed in their work and life. They have no time and patience to listen to me. I am an outspoken person. Children now restrict my activities. (Walking, meeting old friends). They criticize "if you fall ill, we have to look after you. Now medical expenditure is too heavy, we have to bear expenditure of our children". A 76 year old respondent said *"Now I feel I should have joined old age home, it is too late now; it is my fate to be with these people. Till I am hale and healthy all relations work. Now, I am not able to do anything and do not have any income. I am totally deserted".* In the traditional society with their rich experience acquired over the life course, elderly were regarded as repository of social wealth. However, this trend is fast eroding in globalized, modern era. Status of elderly now depends upon the economic contributions they make to the family.

Living Arrangements and Social Support

Living arrangements determine the quality of life of the elderly widows. Sometimes societal and cultural norms forced the elderly to stay with a particular person in the family. Jayashree (2000) pointed out that the pattern of living arrangements has changed due to various socio-economic and demographic factors such as greater geographical mobility, increase in employment opportunities for women, individualism in life and fewer children. In the present study, majority of the respondents were living with their sons. Less than 15% (14.77%) of the respondents stayed with daughters and these respondents either did not have sons or sons were away from them (Table 11.1). Five respondents were living by rotation. Respondents who were living with their relatives were in joint households.

A social network of an individual is vital in the care of elderly. The nature and utility of social support is shaped by the social networks. Social support is always related either to primary or secondary identities. Individual's expressive needs are met in a family, where primary and core relationships are vibrant. Here, family serves as an effective social support system in the routine care, care during ill health and post-hospital care of the elderly. Social support acts as a buffer against illness and sometimes as an ameliorative in disease recovery or care. In Japan, for example, older people who reported a lack of social contact were 1–5 times more likely to die in the next 3 years than were those with higher social support. (Sugiswawa et al. 1994).

This study indicates that primary groups play a vital role in social support system. They were the core social support providers. Widows were emotionally dependent upon their children, especially their daughters for their emotional needs. Overwhelming majority of the respondents depended solely on son for financial and health matters. In Indian society, where there is no universal social security system to cater the needs and security of older person, they depend only on family members for their needs and security. Hence, strong family and community support is needed for their well-being. Informal support systems for widows will be disrupted during old age because of feminization of ageing; with the loss of husband, family members respond and receive her differently. The widow begins to develop skills and abilities or use previously dormant ones to re-pattern life. New goals are set, and life begins to have meaning and purpose again. Gradually, the widow negotiates and re-establishes salient reference groups and social roles into support systems, which meet a variety of needs and require similar reciprocation. These support systems reintegrate the widow into the social system and facilitate the development of a positive identity, a new lifestyle and, often, an altered value system (Hienemann 1980:39). It is clearly evident in the present study that widows were totally adjusted to a new phase of life and not worried about widowhood maladies.

In the present study, 47.72% of the respondents reported that their overall social support network was bad and one-fourth had networks that were neither good nor bad (Table 11.8). The respondents' needs and expectations did not match with that of the younger generation or might not be tolerated and accepted by younger generation. Social support is ensured, but the way it was provided and offered might not be tolerated by the respondents. Of late, even middle-class families struggle with the demands of living, which is becoming costlier day by day; daughters-in-law, who are the traditional caregivers of the elderly, are increasingly taking up the paid jobs for improving the economy of the family. Under such compelling circumstances, it has become stressful for this category of caregivers to devote adequate time and effort, both at work and back at home while taking care of elderly along with other household activities. This has, in turn, led to the weakening of the traditional elderly support system.

Table 11.8 Overall social support network of the respondents

S. No.	Responses	Frequency	Percentage
01	Very good	12	13.63
02	Good	08	9.09
03	Neither good nor bad	22	25.00
04	Bad	42	47.72
05	Very bad	04	4.54
	Total	88	100.00

Health

Self-rated health has been found to be a strong predicator of current and future health (Benjamins and others 2004). A recent survey by Indian Council for Medical Research paints a grim picture of health of older women (DeyA.Wilson and Goel 2010). Illiteracy, depression, cognitive difficulties and problems with activities of daily life (ADL) lowered quality of life of these women. In general, the health of women in post-reproductive years is a low priority for the health sector. For a long time, cardiac problems were seen as "men's problem".

Respondents were asked to rate their general health conditions as perceived by them (poor, fair, very good and excellent) by evaluating last 6 months health conditions. When asked about the overall health status of the respondents, 43.18% of the respondents pointed out that it was "poor" and another 43.18% pointed out that it was "fair" (Table 11.9). Poor health condition of the respondents was attributed to their age, poverty, neglect, lack of caring, lower income and lack of awareness. In India, majority of the women engage in unpaid work due to rampant illiteracy. Due to their low socio-economic status, they have been denied nutritious food, good education and health which lead to ill health. In most of the cases, respondents attributed their ailments to their advancing age.

Disease prevention was totally absent among women with 28.40% of the respondents going for walk either in the morning or in the evening. As majority of them had crossed 70 years, they found it difficult to keep their health intact (Table 11.10).

In the present study, respondents who have very good/excellent health status were practicing yoga, went for regular walk and had their meals on time. Though majority of the respondents were suffering from multiple ailments, joint pains, hypertension and diabetes were bothering them more (Table 11.11). These ailments restricted their mobility and food intake. When all the other family members were enjoying good food, they could not have it due to these ailments. A 69-old

Table 11.9 Present health status

S. No.	Health status	Frequency	Percentage
01	Poor	38	43.18
02	Fair	38	43.18
03	Very good	08	9.10
04	Excellent	04	4.54
	Total	88	100.00

Table 11.10 Practice of yoga/exercise

S. No.	Responses	Yoga		Exercise		Walking	
		Fr	%	Fr	%	Fr	%
01	Yes	4	4.54	–	–	25	28.40
02	No	84	95.45	88	100	63	71.59
	Total	88	100	88	100	88	100

Table 11.11 Ailments bothering you more

S. No.	Ailments	Frequency	Percentage
01	Hypertension	61	69.31
02	Diabetes/sugar	68	77.27
03	Problems related to eye/ear/skin/tongue/nose gynec problems	28	31.81
04	Postoperative problems	27	30.68
05	Joint pains	74	84.09
06	Others	16	18.18

respondent lamented that, "for me there is no festival, all days are same and I have to follow pathya (diet) for whole of my life".

Needs of the Elderly and Policy Issues

The study also attempted to understand the expectations of the respondents. They were asked about their major expectations from the family. Their expectations were classified into three categories, namely physical, financial and emotional/mental. In the physical category, toilets/washrooms inside the house was one of their most important expectations and 44.31% of the respondents expected a separate room (Table 11.12). In the financial category, most of the respondents (85.22%) reported that they need financial independence and all the respondents expected old age/widow pension. Under emotional category, almost all the respondents reported that they expected love and health care, from their family and 86.36% of the respondents expected overall care from the family.

Many of the old-old were frail, some had activity limitations, some were bed-ridden and some with debilitating diseases. Even those who were active found their circle of friends and sphere of activities dwindling. Even under the best of circumstances a family may be always considerate in dispensing care to its old-old kin, and this is more so when he/she is a liability in physical and medical terms. It is found in the study that overwhelming majority of the old-old lived with children and lower number of respondents stayed away because of litigation with their children. Childlessness forced some respondents to stay with their relatives and in joint households.

The research identified three areas where the problem is conspicuous—support system, health care and living arrangement. Since women live longer than men and the extended period of life is spent more or less in morbidity and neglect, it is suggested that the amount of old-age pension should be higher for women than for men and the difference should be maintained and the amount should be increased at ten yearly intervals after attaining 60 years.

Table 11.12 Expectations (N = 88)

S. No.	Expectations	Yes		No	
01	**Physical Facilities**	**Frequency**	**Percentage**	**Frequency**	**Percentage**
	Separate room	39	44.31	28	31.81
	Toilet inside the house	88	100	–	–
	Cupboard	45	51.13	30	34.09
	Servant/Nurse	23	26.13	40	45.45
02	**Financial Facilities**				
	Savings	25	28.40	28	31.81
	Financial independence	75	85.22	–	–
	Pension	88	100	–	–
03	**Emotional and Mental**				
	Love	88	100	–	100
	Proximity with grandchildren	56	63.63	29	32.95
	Companionship	45	51.13	36	40.90
	Independent/freedom	69	78.40	21	23.86
	Overall care	76	86.36	–	–
	Health care	88	100	–	–
	Outings	12	13.63	47	53.40
	Respect	63	71.59	47	53.40
	Moral support	71	80.68	23	26.13

References

Benjamins, M. and others. (2004). Self-reported health and adult mortality risk: An analysis of cause-specific mortality. *Social Science & Medicine, 59,* 1297–1306.

Census of India. (2011). *Provisional population totals.* Registrar General and Census Commissioner of India, Ministry of Home Affairs, New Delhi, India.

Dunlop, B. D. (1980). Expanded home-based care for the impaired elderly: Solution or pipe dream? *American Journal of Public Health, 70*(5), 514–519.

DeyA.Wilson, W., & Goel, A. (2010). Ageing in India: Health and health care issues. In R. Sharma & C. J. Thomas (Eds.), *Ageism. Problems and prospects* (pp. 40–53). New Delhi: Akansha Publishing House.

DiMatteo, M. Robin and Ron Hays. (1981). Social support and serious illness. In B. H. Gottlieb (ed.), *Social Network and Social Support* (pp. 117–148). Beverly Hills, CA: Sage.

Haynes, R. B. (1976). A Critical Review of 'determinants' of patient compliance with therapeutic regimens. In D. L. Sackett & R. B. Haynes (Eds.), *Compliance with therapeutic regimens* (pp. 26–39). Baltimore: The Johns Hopkins University Press.

Hienemann, G. D. (1980). Determinants of primary support system strength among urban widowed women: Does life Stage makes difference? (Ph.D. dissertation, Department of Sociology, University of Illinois at Chicago, 1980).

Jamuna, D. (1999). Elderly care in India. Conflicting tradition and modernatization. *Proceedings of 3rd International Conference of IAHSA: Ageing Societies in a New Millennium, Honolulu, Hawaii* (pp. 33–35).

Jaiprakash, I. (2003). "Impact of out- migration of the young on Older People". *Help Age India Research and Development Journal, 9*(1), January.

Jayashree. (June 2000). Living arrangements and exchange patterns of elderly. *Research and Development Journal 6*(3), 22–27.

Jayashree. (February 2013). ADL and IADL among older people and its impact on longevity: Gender dimensions. *Quarterly Journal of the International Institute on ageing (United Nations-Malta), 23*(2), 18–23. ISSN: 1016-5177.

Knodel, J. E., Chayovan, N., & Siriboon, S. (1992). "The familial support system of Thai elderly: An overview". *Asia Pacific population journal, 7*(3), 105–126.

Lopata, H. Z. (1979). *Women as widows: Support systems*. New York: Elservier North Holland.

Office of the Registrar General and Census Commissioner. (2006). *Population projections for India and States 2001–2026 (Revised December 2006)*.Office of the Registrar General and Census Commissioner, New Delhi.

Shanas. (1979a). The family as a social support system in old age. *The Gerontologist, 19*, 169–174.

Shanas. (1979b). Social myth as hypothis: The case of the family relations of old people. *The Gerontologist, 19*, 3–9.

Sugiswawa, S., Liang, J., & Liu, X. (1994). Social networks, social support and mortality among older people in Japan. *Journal of Gerontology, 49*, S3–S13.

Chapter 12
Caring for the Elderly with Dementia

Robert Mathew

Abstract Dementia is a symptom complex where in memory loss is the main symptom. This is commonly seen in the elderly. Care of patients with dementia is unique and challenging. This is mostly because dementia does not have a cure and takes a protracted course. In addition to supporting the cognitive needs of the patient, caregivers need to adjust to and contain the varied behavioral abnormalities of these patients. Since many medicines used in these frail elderly are more injurious than curative, non-pharmacologic management is the preferred treatment. In this context, caring for elderly patients becomes very important. Through proper care, many of the problems warranting drug administration can be handled without drugs. To achieve this, the caregiver should have a proper understanding of the disease. Since taking care of dementia is a difficult task, physical and mental health of the caregiver also has to be taken care of.

Keywords Dementia · Caregiver · Pharmacologic management · Non-pharmacologic management · Day care center · Thiruvananthapuram model

Introduction

Dementia is a symptom complex in which memory loss is the main symptom. This is essentially a disease of the elderly even though it can occasionally occur in the young. In addition to memory loss, there will be one or more other mental faculties (cognitive domains) involved. These may include ability to orient oneself with the space around and people around. They will have problems recognizing the time of the day or the date. Other cognitive domains involved include language functions, calculation-related functions, ability to construct things, and ability to execute complex commands. To start with, these impairments may be mild, but as time progress, these symptoms aggravate and even day-to-day life becomes difficult. In

R. Mathew (✉)
Mookambika Institute for Medical Sciences, Kulasekharam, Tamil Nadu, India
e-mail: robert_mathew90@yahoo.co.in; robertmathew90@gmail.com

S. Irudaya Rajan and G. Balagopal (eds.), *Elderly Care in India*,
DOI 10.1007/978-981-10-3439-8_12

due course, people with dementia develop behavioral abnormalities mostly in the form of becoming depressed, apathetic, irritable, and losing the drive to function. They may develop florid features of psychiatric illness like suspiciousness and tendency to abuse others. Some develop abnormalities in mood either becoming excessively depressed or becoming excessively elated. They also develop false thoughts (delusions), false perceptions (hallucinations), or misperception (illusion). As time goes on, they develop other neurologic problems in addition to cognitive abnormalities. They may lose control of their bladder and bowel, thereby resulting in incontinence of urine or feces. The patients also develop physical disability in the form of walking difficulty and swallowing difficulty during advanced stage of disease. The commonest type of dementia is called Alzheimer's dementia (AD), and most of the clinical features of dementia are described based on this disease.

Care of patients with dementia is unique in many ways. Dementia occurs in the elderly, and usually, they will have other comorbid conditions which need to be taken care of. The elderly age group are more sensitive to medicines and for the same reason more prone to develop medicine-related adverse reactions. Dementia in most of the patients is degenerative which means that it does not have a cure and is gradually progressive. In the absence of curative treatment, proper care is all that can be offered. The long protracted course of the disease means that it warrants prolonged care from caregiver. This adds to the disease load in society. In a good number of people with dementia, the primary caregiver will be the spouse of the patient who is equally or more aged with associated comorbid conditions.

Taking care of this group of elderly patients with a disease that is not curable and which is gradually progressive with a protracted course is a formidable task and is the subject of discussion in this chapter. The persons who take care of patients with dementia themselves do have their own need for care. Persons who live with and provide care for patients with Alzheimer's disease, often report emotional stress due to caring for the dementia patients. Also, this is partly caused by other issues like the need to give up vacations and avoid hobbies as part of patient care. As a result, care givers need support and proper education as regards the disease (Mayeux 2010).

Diagnosis of Dementia

Care of patients with dementia starts with the diagnosis of the illness. It needs a specialist in the form of a psychiatrist or neurologist or geriatrician to confidently diagnose dementia. Most of the dementing illnesses do not have a single test by which the disease can be conclusively diagnosed. In most of the instances, a diagnosis is made based on the clinician's bedside examination, supplemented by some essential investigations. The most important investigation for early diagnosis and characterization of dementia is neuropsychologic evaluation. This is carried out by a neuropsychologist with special training in memory disorders. The evaluation consists essentially of making the patient perform tasks which involves various cognitive domains of which memory is one. For example, the patient is asked to remember three words and recollect them after 5 min. The response that the patient gives is compared with the response given by normal people of that age and education. The extent to which each brain function is involved is thus quantified and final conclusion made as regards presence or absence of dementia. The test also gives information as regards the extent of involvement of various intellectual functions (cognitive functions) of the brain which is helpful in diagnosing the underlying dementia subtype.

Dementia by and large is considered to be a disease without curative treatment. However, 10–20% of patients with dementia may have a curative treatment. Having diagnosed that a patient has dementia, the next step is to find out whether the patient has a curable cause for the dementing illness. This may need some investigations of which probably the most important one is the imaging of brain. This investigation is called brain scanning. This is a complex investigation which involves generating pictures of the patient's brain using different computer-based techniques. The pictures thus taken may give an idea of the structure of the patient's brain as well as function of various regions of the brain. This is supplemented by some blood tests as well. At the end of all these investigations, conclusion can be drawn as regards curability of the disease. The vast majority of dementias which are not curable are categorized as degenerative dementias, thereby indicating two things: 1. the disease is not curable and 2. the disease is gradually progressive.

Drug Treatment of Dementia

Before discussing the non-pharmacologic care given to dementia patients, which probably is the most important modality of care, brief mention should be made about drug treatment of these patients. Drugs do play an important role in the management of patients with dementia. At the moment, it is considered that in more than 80% of patients, the dementia is due to diseases which are nor curable. However, there are drugs which give symptomatic relief of the memory loss. Also there are drugs available for various behavioral symptoms which these patients develop during the course of illness. Many of the drugs may have to be given for a prolonged period of time. Selecting the appropriate drug and regularly evaluating the response to the drugs are important in drug management of dementia. In general, the number of drugs used should be minimized to reduce the adverse effects as well as to prevent drug-to-drug interactions. Elderly population in general are excessively sensitive to drug effects, as well as side effects, and hence, most of the drugs need to be started at a small dose and gradually titrated to the desirable dose level. Other than drug-related side effects and drug-to-drug interaction, there may be patient-related problems like adherence problems or inappropriate use of drug by the patient. Inappropriate drug selection can also rarely lead to unwanted side effects. This is all the more important in patients with dementia. One simple example is anticholinergic effect. This is an undesirable side effect of some of the drugs especially of some drugs used for behavioral modification as well as for

Parkinsonism. In otherwise healthy candidates, these are very useful and would give good symptomatic relief. But in patients with dementia, the anticholinergic side effects will worsen the dementia symptoms. Patients may get confused and disoriented as a side effect of the drug. So, these medicines are inappropriate for dementia patients. Multiple drug usage is a common co-factor for drug-related side effects and can often lead to worsening of cognitive status. It may lead to further reduction in capability of daily activities. Even if the drug selection is appropriate, each additionally given drugs increase the risk of clinically relevant drug-to-drug interaction. These drug-to-drug interactions can produce unwanted side effects leading to hospitalization, increased risk of death, and decreased quality of life.

Drugs inducing sleep (sedative drugs) can increase the risk of falls and hip fractures. Poly pharmacy leads to poor compliance by the patient and increases the cost of care. Some elderly are in the habit of taking off-label drugs, i.e., taking drugs not really prescribed by doctor. This may lead on to drug abuse and related side effects. These problems may get aggravated if the patient gets demented. Hence, a comprehensive review of medicine intake by the patient is always advisable and will certainly improve the clinical situation of the patient (Fiß et al. 2013).

Pneumonia, febrile episodes, and eating problems (with consequent salt and water depletion and other nutrition-related complications) are frequent in patients with advanced dementia, and these complications are associated with high 6-month mortality rates. These problems are more related to general medical health than the underlying dementing illness. Nevertheless, these problems also need medical management with drugs. Distressing symptoms and burdensome interventions are also common among such patients, especially so in the advanced stage of disease. Whether such over aggressive interventions for such a disease in its advanced stage will have any bearing on the improved outcome in general or at least quality of life is a debatable issue (Mitchell et al. 2009).

None of the degenerative dementias have any curative treatment. In the management of these patients, the memory loss and other cognitive abnormalities that it produces are of a lighter degree. The major problems in management of dementia patients however arise from the various behavioral abnormalities these patients have. Hence, the most important part in the management of patients with dementia is probably the management of the various behavioral abnormalities associated with the same. Non-pharmacologic management forms a very important part in management of various behavioral abnormalities in patients with dementia.

The commonly seen behavioral abnormalities in dementia patients are (Overshott and Burns 2005):

- Apathy
- Aggression
- Depression
- Agitation
- Wandering
- Delusion
- Hallucinations

- Kicking
- Restlessness
- Screaming
- Shaking
- Poor sleep
- Biting

Treatment: Non-Pharmacologic

Non-pharmacologic treatment is very important and often overlooked in dementia care. Elderly patients are excessively sensitive to medicines and hence are best managed without medicines as far as possible. Non-pharmacologic management has no risk of adverse events and/or drug interaction. Also it should be remembered that environmental factors and interpersonal factors are frequent triggers of behavioral problems, which can be tackled largely with non-pharmacologic management. Environmental factors include false or improper lighting, noisy surroundings, and frequent change in place of stay. Interpersonal factors include unnecessary argument with patient and intimidating body language. Attention to these problems with no pharmacologic approaches can be quite effective in improving the situation and alleviating behavioral problem.

Providing education about the disease as well as supporting the caregiver is important. The caregiver should be given advice about managing patients with dementia and coping with the various problems arising thereof. This forms the core of dementia care as well as treatment. Modification of the environment as well as behaviors of caregivers becomes extremely important in caring of patients with dementia. The strategies may be different from patient to patient and for the same person from time to time.

The various non-pharmacologic management strategies being tried out in patients with dementia are as follows:

Verbal therapies

- Behavioral therapy
- Reminiscence therapy
- Cognitive behavioral therapy
- Interpersonal therapy
- Reality orientation
- Validation therapy

Non-verbal therapies

- Music therapy
- Aromatherapy
- Exercise and activities

- Bright light therapy
- Multisensory therapy

A detailed discussion of all of these strategies is beyond the scope of this chapter. However, discussion of the important behavioral abnormalities and relevant non-pharmacologic management of the same are included.

Clinical case 1: Mrs. X, 72 years, was diagnosed as having Alzheimer's dementia 10 years ago. Ten years into the illness, she was dependent on others for all activities of daily living. She would often misidentify even relatives as close as her own children. One day, she started to search for her second son who was working abroad. Attempts to convince her about his absence at home failed. She became increasingly irritable and refused to take food. Family members made the son abroad to talk to her and convince her, which did not convince her. The elder son who was staying with her tried another strategy. He told her that he was the son who lived abroad and had returned recently. This worked and she became very happy to see the son who had returned. Her irritability also settled down and she became much easy to manage.

Lot of literature is available on these types of coping strategies for patients with dementia. These include booklets, books, and videotapes which educates caregivers on various strategies. There are significant number of papers comparing non-pharmacologic management and pharmacologic management. Studies say that non-pharmacologic management can significantly reduce behavioral problems associated with dementia, reduce the incidence of depression in patients and delay institutionalization of dementia patients.

ABC of Behavioral Modification

Caregiver needs to know about the ABC of behavioral changes. A is the trigger or the antecedent event of the behavior change, B is the abnormal behavior that has to be tackled, and C is the consequence of that behavior. Once the caregiver understands the ABC of the behavior change, he/she can avert it either by intervening at the level of A, managing the triggering event or managing the adverse consequence of the behavior (Ross and Bowen 2002).

Case 2: Seventy-four-year-old Mr X with advanced dementia started shouting and yelling one day in the afternoon. He was fairly comfortable in the morning hours, but by evening, he became agitated, angry, restless, and started to shout. Trying to pacify him did not work at all (intervention at level C). In the evening, he was found to have fever and pain on passing urine. It was the pain of urination and fever that triggered his aggressive behavior. He was given medicines for urinary tract infection and fever. He became calm and composed over the next three days (intervention at level A). Once dementia patients develop medical illness not directly related to the brain, they become either excessively less active, or get agitated, disoriented, and even aggressive. Some patients may develop incontinence of urine along with this. This is a specific medical condition called Delirium. So, our patient had developed delirium consequent to the urinary tract infection. He improved on successful treatment of the underlying medical problem (Urinary tract infection and fever).

For each and every behavioral problem, the caregiver has to find out the ABC. Medical illness is a well-known trigger for behavioral problems. However, many problems may have more than one antecedent, all of which have to be identified for formulating proper preventive action. Identification of the problem is very important in the A-B-C approach. The more precisely and clearly problems are defined, the more likely that a proper solution can be worked out. The caregiver should also be able to describe it in such a way that others will get a holistic picture of the problem. Let us take a simple example of the patient "arguing unnecessarily". When does the patient argue? Is there any specific situation, or persons triggering the argument? Or is it that the patient keeps on arguing whatever be the situation? Is the argument with a reason or totally irrational? Is the caregiver able to give specific examples of argument? The more details the caregiver could collect about the problem, the more likely that he will find an answer for the same.

The next step is to set realistic goals. The goals should be precise and achievable. It should be tailored to the patient and care giver. For example, the aim of the caregiver is to keep the patient happy always. But, this may be too difficult to attain. However, smaller achievable goals like "increasing pleasant activities" or "avoiding discussion of upsetting topics" are more feasible. Goals should be small and achievable to start with, and should be gradually increased to more difficult ones. Patients as well as caregivers should be rewarded as an encouragement for further achievement. The targets and plans should be continually evaluated and changes should be made. The plans should be carried out consistently, and at the same time, the planner should be flexible enough to modify it dependent upon the requirement of the patient and caregiver.

In the A-B-C approach, the caregiver is encouraged to form their own strategies to manage situations. A brief discussion of some of the common behavioral problems and approach to their management is as follows.

Depression

Depression is very common in dementia. Depression occurs in 30–70% of patients with dementia. Unhappiness, inactivity, withdrawal, tearfulness, fatigue, expressions of guilt and worthlessness, difficulty to concentrate, excessive sleep or loss of sleep, excessive appetite or loss of appetite, and loss of interest in day-to-day activities are all features of depression. Patients lose their ability to get involved in pleasurable activities, and they may disengage from all activities altogether. Certain interventions can modify the situation. The enjoyable activities which the patient can perform should be encouraged and made enjoyable. Probing should be done into the activities that the patients enjoyed earlier, and such activities should be brought back as far as possible. They should be modified appropriately so as to fit with the current physical and mental disabilities. These may include watching movies on television, listening to music, going out for a walk, having meals with family or friends, doing exercise, going for a ride in a car, and helping in household activities.

Attempts should be made to redirect the focus of the patient. Depressed patients often get obsessed with depressive thoughts and get more and more depressed. They should be distracted from these thoughts. These thoughts should be replaced by non-depressing thoughts. Viewing photographs of the happier prior years, reminiscing of pleasant experiences, having cheerful conversations etc. can distract them from depressing thoughts.

Increasing social activity is another way of reducing depression. Activities that can keep the patient alert and involved can improve depression significantly. Social visits, outing with persons whose company the patient enjoys, and encouraging visits by loved relatives are all means of elevating the mood.

Elimination of sources of frustration and conflict also improves mood. If the patient keeps on attempting tasks which they cannot really do, it will lead on to more frustration and depression. Continually challenging the patient's capabilities especially those which are already compromised leads to onset of depression or worsening of existing depression. Sources of conflict and frustration should be identified and eliminated as far as possible.

Mood of the caregiver has an important bearing on the mood of the patient. Depressed caregiver often adds to the depression of the patient. Caregivers should be occasionally screened for depression, and measures should be taken to improve the caregiver's mood. This will certainly have a positive outcome on the patient's behavioral problems.

Agitation

Agitation is another common behavioral abnormality in patients with dementia. Agitation includes irritability, restlessness, physical and verbal aggression, non-cooperation or refusal of personal care assistance, pacing, and wandering. Agitation is expected to affect between 30 and 70% of dementia patients sometimes during the course of illness. Agitation adds to the caregiver's burden tremendously and can turn injurious to the patient and others. Agitation can be triggered by psychological causes in addition to environmental and interpersonal causes. These causes may interact with each other as well. For example, patient may become less tolerant to frustration due to advancement of the disease. Without sensing this, the caregiver may keep on challenging the patient's memory by giving tasks which could have been avoided. The patient will fail in these tasks and get more and more frustrated. This coupled with poor frustration tolerance, which are part of the disease process, and may lead to catastrophic reaction. However, if the caregiver becomes more supportive and avoids too much of challenges, the agitation of the patient improves. So in handling agitation, it is very important that the caregivers identify triggers and avoid them as far as possible. It is equally important that the triggers should be identified earlier and intervention made before extensive damage occurs or catastrophic reaction occurs.

The caregiver should remain calm if he wishes to calm the patient. The voice of the caregiver as well as body language should be reassuring and gentle. Touching of the patient should be done judiciously and gently. Necessary steps should be taken to make sure that the patient does not perceive it as a threat. Arguing with these patients or even trying to reason out with an agitated person may often lead to worsening of the situation, rather than clarifying the issues. It should be remembered that the demented patient will have problem in following and understanding long sentences, complex sentences, double negatives, puns etc., and such complex language forms are better avoided during conversations.

An agitated patient is always best approached slowly and that too from the front, as far as possible. Care should be taken to avoid startling the patient by approaching from the side or behind abruptly. It is better to stand or sit at the eye level than standing above the patient.

Distracting the patient may help in some occasions. Patient may be distracted by asking questions. The initial questions may be about the problem itself as patient will be cooperative and listen to such questions. But subsequently, related and unrelated topics can be brought into discussion and gradually, patients attention can be shifted away from the burning issue to something cool and comforting. Patient may be gently taken away from the activity he was already doing, and if needed changed from the present room or location.

In spite of these attempts if the patient continues to be agitated and threatens with injurious behavior, it is always better to get help of more a professional to prevent or contain damage. In spite of all this, there may be occasions where drugs for calming down the patient may have to be used.

Wandering

Wandering is another behavioral abnormality which needs consideration, as it can turn out injurious to the patient and create profound problems for the caregivers and family. Wandering tendency has been reported in up to 65% of patients and is usually seen in advanced stages of the illness. Wandering is found to be related to the severity of cognitive impairment and the rate of cognitive decline. It may even indicate decreased length of survival. However, wandering can occur at all levels of cognitive impairment. Caregiver distress increases significantly with wandering tendency. Wandering can be prevented by environmental modification, using electronic alarm system, by reinforcing non-wandering behavior, and by keeping the patient otherwise engaged and active.

A brief discussion of the above-mentioned strategies follows:

1. Environmental modification is mostly in the form of providing visual cues with a view of keeping the patient well oriented in place. Sign boards and labels do

help a lot in prevention of wandering. Important rooms like bathrooms and bedrooms can be labeled well and provided with good sign boards. Visual cues can be used to prevent the patient from wandering into dangerous and potentially injurious situations. Stop signs can be used if the patient needs to be prevented from entering into rooms like the kitchen. Cues like coat, hat, umbrella etc. used for moving out of house should be removed from the scene, as they may trigger desire to wander out. Disguising exit doors by blending them with the wall or by hiding them with curtains may to some extent prevent the patient from banging on the door. Doors may have to be eventually kept locked so that an attempt to open them by the patient fails, and in some situations, this may dissuade them from wandering. The sign boards and visual cues may also help the patient to get back to their pathway once they wander away accidentally. Interesting displays or items may distract the patient and keep them engaged thereby preventing wandering.

2. Security systems will help in minimizing problems created by wandering. Electronic alarm systems will alert the caregivers when the patient strays out of the way. These may include simple ones like those that give an alarm once the exiting door is opened or trespassed, to radio transmitter devices that track the patient wherever they move around. Close-circuit television cameras also help in monitoring the patients.

A safety plan should be charted out. A current photograph of the patient should be kept ready for search operations should the patient wander away. Patient should wear a bracelet that contains necessary information for tracing the patient, in addition to clearly mentioning that the patient has a memory disorder. It should be secured well so that patient will not be able to remove it easily or it may not get lost easily. Labels with necessary information may be attached to clothes as well.

3. Patient should be kept engaged by various interesting activities. Boredom or loneliness is one of the reasons for wandering. Recreational activities, social visits, and other social activities may help in preventing wandering.

Sleep Disturbances

Sleep disturbances is another major behavioral abnormality which is a cause of concern. Patients with dementia tend to sleep more during the daytime and less during the night, which is termed as reversal of sleep rhythm. They have a tendency to wander at night and may get agitated during bedtime. Their sleep also gets disturbed at night. These sleep-related problems coupled with poor sense of time may create considerable problems to the caregivers. There are instances in which patients wake up in the middle of the night thinking that it is early morning and start dressing up to go for their morning walk or to the temple. They tend to disrupt the whole routine of the family by disturbing others' sleep as well. The caregiver is forced to get up several times during night because of these behavioral abnormalities eventually leading to their exhaustion. This exhaustion coupled with failing emotional and physical health often forces the relatives to institutionalize the patient.

Sedatives are not the first-line response to these types of problems as the sedatives by themselves may produce several of unwanted effects. They may lead on to more falls, and they may further worsen the existing memory problem. In addition, these drugs may produce sleep disturbances and important medical problems like aggravation of preexisting medical diseases like obstructive sleep apnea. Obstructive sleep apnea (OSA) is a situation wherein during sleep, proper breathing is interfered with because of obstruction to the airway necessary for breathing. As a result of this obstruction in breathing, the amount of oxygen in blood falls down to undesirably low levels, and the patient is forced to get up to restore proper breathing and replenish blood oxygen. The awakenings happen at a subclinical level, and hence, the patient does not completely become awake, even though the depth of sleep is significantly compromised. These are called micro-awakenings. These micro-awakenings prevent the patient from getting refreshening sleep. As a result, the patient becomes excessively sleepy during the daytime. These are some of the reasons why sedatives are avoided in these patients.

Environmental and behavioral factors do play a role in the sleep-related behavioral abnormalities and need to be tackled for symptomatic relief. Continence practices also contribute to disturbed sleep. Presently, there is a suggestion that timed bright light exposure may reduce the sleep disturbances found in Alzheimer's dementia (AD). Some of the methods useful for non-pharmacologic management of sleep disorders are mentioned herewith.

1. Try to have a consistent bedtime and rising time. Regular sleep-time habits may reinforce the time cues of dementia patients who are already having problems with orientation in time. This may also provide clues to maintain a regular circadian rhythm of the brain. The day–night rhythm is called circadian rhythm. A rhythm city of the brain circadian rhythm may reinforce brains' time-keeping system. Proper sleep timing is considered as one of the important steps in the management of sleep-related problems in all older adults.

2. Reducing daytime napping is a very important part of management of sleep-related problems. As dementia advances, many patients spend substantial amount of time sleeping during daytime. By reducing the daytime sleep and by engaging in some form of physical activities during daytime, patients tend to get a longer and more consolidated sleep during night.

3. Dietary modification can help to some extent. Restring the use of alcohol- and caffeine-containing beverages during bedtime may improve nighttime sleep. These food items are well-known stimulants that are capable of disturbing nighttime sleep. Foods like banana, cheese, and milk are supposed to promote nighttime sleep. Light snack with such products at bedtime will facilitate sleep and prevent nighttime awakening due to hunger.

4. Environmental disturbances should be minimized. Nighttime light level and noise levels should be minimized as far as possible. Other nighttime environmental disturbances to be taken care of include disturbances caused by household pets, grand children, other adults at home, caregiver snoring, and vehicles.

5. Change in regular routine such as family visits, holidays, or long-distance travel is likely to aggravate sleep disturbance. Hence when such activities are planned, allowance should be given for catch-up sleep.

Role of the Caregiver

Caregiver has an important role to play in the caring of patients with dementia. Caregivers are the key persons in executing various non-pharmacologic management techniques. Caregiver has a very unique role in identifying and defining various problems of the patients and formulating appropriate behavioral and environmental strategies to counter those behavioral problems. Hence, caregiver education is a very important part of dementia patient care. Caregivers may include spouse, unpaid family members, or paid caregivers. In planning management strategies, we should have a clear awareness of the skill knowledge, energy, and resources of these caregivers. Various factors influencing the caregiver performance also have to be reviewed. For example, the degree of stress the caregiver is experiencing determines how much he/she is willing to carry out a behavioral change plan. Some caregivers will have their own family problems, some will have physical ailments, some may be depressed, all of which ultimately contributes to the execution of the planned strategy. Hence, healthcare providers should spend time with the caregivers to identify their problems and suggest realistic solutions and realistic goals. Treatment strategies should be tailored to meet the needs of the caregiver as well as the patient. As dementia progresses, more and more problems with different dimensions will crop up. Also different dementia types have different behavioral abnormalities. Some types may have lot of false beliefs and thoughts, while in some other types, the main problem may be lack of concern for others emotions. These have to be identified, and strategies have to be modified. Some amount of flexibility is needed in this respect. It should be remembered that the dignity of the patient as well as the caregiver should be respected when strategies are planned.

Problems of Caregiving

Caring for a patient with dementia is a really challenging assignment. In dementia, it is not the memory alone that is lost, but problems with judgment, orientation, and communication. These will increase the work of the caregiver. More distressing will probably be issues related to change in personality of the patient. Some patients become excessively irritated and agitated while others display apathy. Emotionally, caregivers may feel unhappy that none of their caring work is being acknowledged by the patient, because of apathy of the patient. Patients with delusions and hallucinations may turn suspicious of their caregiver and in the bargain make life hell for themselves and the whole family. They may misplace and loose valuable items like ornaments. However, their false belief will make them believe that their articles have been hidden by the caregiver and hence will accuse the care giver of the same. As the disease progresses, patients need more care and attention and hence many caregivers feel isolated. They will have high level of stress and hence problems with employment, income, and consequently have poor financial security. Some of them may develop health-related problems. If the caregiver is in the earning age, he/she will be forced to give up the job and stay back at home for caring, which further complicates the situation. It is true that the caregivers derive some emotional satisfaction from caring; nevertheless, majority report high degree and depression. Increased caregiver stress leads to nursing home placement. Western studies say that majority of family members with dementia report no guilt in deciding to place the patient in a nursing home. Culturally, Indians are different from the Western world and majority of the caregivers resist the idea of nursing home placement or placement in dementia home. But ultimately, especially as the disease advances, many of them finally opt for dementia care home placement. Nursing homes or dementia care centers catering to patients with dementia are extremely uncommon in our country mostly consequent to the lack of awareness of its necessity, though of late, there is an increase in dementia homes (Luchsinger et al. 2012).

Adaptation of Available Experience and Recommendations into Indian Scenario

Vast majority of the literature on dementia at the moment is available from studies conducted in the Western world. Many of these findings and recommendations are not applicable to our patients and populations because of the major differences in demographic variables like education, culture, socioeconomic status etc. In the absence of robust indigenous research data, adaptation of the existing recommendations will be an immediate and feasible early solution. Even conducting a dementia survey in our population of uneducated elderly has lots of hurdles (Mathew and Mathuranath 2008). The various tools available for assessment as well as various management strategies adopted from the west needs modification, standardization, and validation before they can be adapted to our population. Even this is a major task. Some very preliminary work has been done in this direction and lot more has to be done. Only then will our studies will become comparable with that from the west (Mathuranath et al. 2004, 2007). Development of indigenous data and strategies nevertheless stands to be the final answer to all these problems.

Dementia Care: Indian Scenario

India has a huge burden of people with dementia. At the moment, India is estimated to have 3.7 million dementia patients and the number is expected to grow exponentially in years to come. In the Indian scenario, the family is the primary support for the demented person. The dementia patient continues to live with their families. Dementia care is usually a joint effort by the adult members of the family who stay in the same household. It is also common that one person among them would be more involved with the care at any given point of time and that person can be considered as the primary caregiver. The care arrangements in the country are unique, and service development needs to take advantage of the same. The best way will be to improve caregiver support and make use of locally available resources to address the need.

At the moment, there are only half-a-dozen residential care facilities exclusively for people with dementia in whole of India. Around ten day care centers offer professional care for people with dementia. Domiciliary care is provided in around five centers. It is evident that there is a huge service gap, and meeting all the challenges in closing the service gap warrants a multipronged approach.

There is definite evidence from research carried out in India that locally available non-specialist volunteers can be trained in detecting cases and providing effective non-pharmacological interventions for the elders with dementia. In practice, such interventions will need to be incorporated into horizontally constructed programmes addressing the generic needs of frail, dependent older people and their caregivers, whether arising from cognitive, mental, or physical disorders. Good-quality training, research, and strong political commitment to the cause are central to achieving this goal (Dementia India Report 2010).

Dementia Care Center: The Thiruvananthapuram Model

Thiruvananthapuram city is the capital city of Kerala, a small state in the southern most part of Kerala. For the last 10 years, a dementia care center has been successfully run by the Thiruvananthapuram chapter of Alzheimer and Related Disorders India (ARDSI) a non-profit NGO. The center is run in a rented house. Patients with dementia were housed in cubicles of two or three beds. Up to 15 patients can be housed at a time. A social worker with a formal training in medical social work stays in the home as the manager and two or three staff members will be on duty at any point of time. These staff members are volunteers without any formal training in patient care and they work on an honorarium. They are given non-formal training on caring for dementia patients by experts in dementia. Patients are examined by a dementia expert at the time of entry to make sure that there are no major life-threatening situations. Majority of these patients do not have any regular medical checkups let alone dementia evaluation. This is either because of poor access to the facility or because caregivers are not readily available to take them to the facility. The severity of the disease at the time of admission was mostly moderate or severe. Reason for institutionalization was mostly social factors, the commonest of which was absence of proper caregiver. The median duration of stay was 4 years. Reason for discontinuation of stay was mostly medical problems and, rarely, social problems.

Two patients died during the stay. Relatives were encouraged to visit the patients as frequently as possible, and on an average, the relatives made at least two visits in a year. Most of the caregivers were satisfied with the stay and living conditions.

The caregivers of the center did have problems handling various behavioral problems of the inmates and had consequently developed anxiety. However, on the whole, their mental health was good and none of them had any stress-related problems warranting medications. With proper guidance and advice, the caregivers mastered skills by which they could tackle many of the behavioral problems. Many of the behavioral problems which the family caregivers found difficult to manage were well tackled in the care center. However, most of the patients were on regular dose of neuroleptics which were started earlier (neuroleptics are drugs which modify the various behavioral abnormalities of the patients including agitation). Attempt at reducing the dose of these drugs during stay at the center failed in most cases.

The caregivers would pay for the service on a monthly basis. The monthly payment was less than the amount needed to maintain professionals at home. So, stay at the care center was economically viable. This model suggests that it is possible to run a dementia care center successfully without dependence on professional health workers or specialists.

References

Dementia India Report. (2010). Retrieved July 07, 2016, from http://ardsi.org/downloads/main%20report.pdf.

Fiß, T., Thyrian, J. R., Wucherer, D., Aßmann, G., Kilimann, I., Teipel, S. J., et al. (2013). Medication management for people with dementia in primary care: Description of implementation in the DelpHi study. *BMC Geriatrics, 13*(13), 121.

Luchsinger, J., Mittelman, M., Mejia, M., Silver, S., Lucero, R. J., Ramirez, M., Kong, J., Teresi, J. A. (2012). The Northern Manhattan caregiver intervention project: A randomised trial testing the effectiveness of a dementia caregiver intervention in Hispanics in New York City. BMJ Open. 2012 Sep 14;2(5). pii: e001941. doi:10.1136/bmjopen-2012-001941. Print 2012.

Mathew, R., Mathuranath, P. S. (2008). Issues in evaluation of cognition in the elderly in developing countries. *Annals of Indian Academy of Neurology, 11*(2), 82–88.

Mathuranath, P, S., Hodges, J. R., Mathew, R., Cherian, P. J., George, A., Bak, T. H. (2004). Adaptation of the ACE for a Malayalam speaking population in Southern India. Adaptation of the ACE (and the MMSE) for an Indian population. *International Journal of Geriatric Psychiatry, 19*(12), 1188–1194.

Mathuranath, P. S., Cherian, J. P., Mathew, R., George, A., Alexander, A., & Sarma, S. P. (2007). Mini mental state examination and the addenbrooke's cognitive examination: Effect of education and norms for a multicultural population. *Neurology India, 55*(2), 106–110.

Mayeux, R. (2010). Clinical practice. Early Alzheimer's disease. *New England Journal of Medicine, 362*(23), 2194–2201.

Mitchell, S. L., Teno, J. M., Kiely, D. K., Shaffer, M. L., Jones, R. N., Prigerson, H. G., et al. (2009). The clinical course of advanced dementia. *New England Journal of Medicine, 361*(16), 1529–1538.

Overshott, R., & Burns, A. (2005). Treatment of dementia. *Journal of Neurology, Neurosurgery and Psychiatry, 76*(Suppl 5), v53–v59.

Ross, G. W., & Bowen, J. D. (2002). The diagnosis and differential diagnosis of dementia. *Medicine Clinics of North America, 86*(3), 455–476.

Chapter 13
Care and Support Arrangements Among Elderly Residents of an Urban Slum in Tamil Nadu State, India

Gayathri Balagopal

Abstract For the socio-economically disadvantaged elderly who have aged with multiple deprivations, the onset of old age is marked by ill health, difficulties in functioning for some, coupled with decline in their already meagre economic resources and changes in their family composition (like loss of spouse). The ability of elderly to cope with old age and illness depends not just on their functional and financial independence, but also on care and support that they receive from family, state, non-governmental organisations and private-for-profit sector. This research investigates caregiving arrangements among the elderly based on primary data from a field survey conducted in 2005 in an urban slum in Chennai. The findings reveal that though co-residence rates with immediate family members and frequent face-to-face contact with non-resident children were high among elderly respondents, it did not translate into high social care provision for the elderly, as most were able to carry out ADL independently even if with some difficulty. Given the fragile economic condition of the elderly, majority of them received economic support. Among those who received care and support, family involvement was substantial, and the role of the state was significant in economic support (social security pensions) but absent in social care provision. Social care displayed gender asymmetry, with women providing bulk of care to the elderly, whereas most of the care recipients were elderly men. However, elderly women seem to be better resourced in terms of social networks to deal with old age, as they received care from diverse sources, unlike elderly men, whose predominant source of social care was their spouse. The gender asymmetry in caregiving has to be redistributed among other family members and the state, with a need to plan for formal, home-based care mechanisms for the elderly, in addition to universalisation of social security pensions for the elderly.

Keywords Elderly · Slum · Morbidity · Caregiving · Escort to health facility · Care at home · Economic support · Caregiver

---

G. Balagopal (✉)
D-10 Adhisri, New # 16 Kothari Road, Nungambakkam, Chennai 600034, Tamil Nadu, India
e-mail: balagopalgayathri@gmail.com

S. Irudaya Rajan and G. Balagopal (eds.), *Elderly Care in India*,
DOI 10.1007/978-981-10-3439-8_13

Introduction

In 2001, while the elderly (76.6 million) constituted 7.5% of the total population of India, by 2011, they (103.8 million) comprised 8.6% of the total population (Government of India 2016). Projections for 2051 by Irudaya Rajan and Mathew (2008) reveal that the share of the elderly population will be 17.3%. According to the Census of India in 2001, among the major states, the proportion of elderly were highest in Kerala (10.5%), Punjab (9%) and Tamil Nadu (8.8%). In 2011, Kerala (12.6%), Tamil Nadu (10.4%) and Punjab (10.3%) had the highest proportion of elderly among the major states (Government of India 2016). Many of the issues that confront the elderly in India have their origins in living conditions before commencement of old age, which in turn are a result of poor promotional and preventive social security. For the socio-economically disadvantaged elderly who have aged with multiple deprivations, the onset of old age is marked by ill health, difficulties in functioning for some, coupled with decline in their already meagre economic resources and changes in their family composition (like loss of spouse). The ability of elderly to cope with old age and illness depends not just on their functional and financial independence, but also on institutions that comprise the care diamond (Razavi 2007), namely, family, state non-governmental organisations (NGOs) and private-for-profit sector. In India, policy and discourse have deified the gendered nature of care provision, with females expected to bear the burden of unpaid care work (Palriwala and Neetha 2011). Consequently, care for elderly, persons with disability or ill health and children, when provided, are mostly by women, given the absence of state involvement and inability to pay for home-based market-provided care. Economic dependence is rooted in past involvement in the low-paid informal sector, poor access to occupational pension, low access to employment among women and inadequate accumulation of assets. This dependency is further exacerbated after the onset of old age by out-of-pocket expenditure on illness and absence of universal social security. The central role that family plays in elder care, which is perpetuated by policy, is prevalent in other Asian countries like Korea and Japan. In Korea, policy had devolved care responsibilities to the household, which means to female family members, although in the recent past, the country has witnessed a shift from the extensive familialism to a form of modified familialism through expansion of social care provision by the state (Peng 2009). In Japan, bulk of caregiving for the elderly is provided by female family members (Abe 2010). Eeuwijk (2006) emphasises the dominance of family in care provision for the elderly and the gendered nature of caregiving, with wives, daughters and daughters-in-law providing most of the care in Indonesia. Social networks are considered to be potential sources of social support and women have more diverse networks like kin-based, neighbours, friends, siblings. when compared to men (Grundy 2006). The reliance on families to provide almost all care needs of the elderly results in caregivers experiencing burden, which Eeuwijk (2006) has classified into physical burden, economic burden, social burden, psychological burden and infrastructure burden. Physical burden include tasks like assisting the elderly with toileting, bathing;

economic burden is a result of expenditure on health care, special food, assistive devices on the elderly; social burdens refer to inter-personal conflicts between the elderly and their caregiver; psychological burden happens when caregivers find it difficult to continue with long-term physically demanding care; and infrastructure burden refers to restricted housing space, water supply, etc. (Eeuwijk 2006).

Lowenstein and Daatland (2006) have identified four approaches in studies on inter-generational solidarity, namely retrospective theories, situational influences, prospective theories and a combination of all three. Retrospective theories examine the role of cultural practices and normative beliefs; situational approach takes into account factors like competing obligations and inter-personal relationships that promote or impede care; prospective theories tries to understand future consequences of present actions; and a mixed conceptual approach which explores past, present and future influences together (Lowenstein and Daatland 2006). In a cross-country comparison (Norway, UK, Germany, Spain and Israel), it emerged that although familialistic attitudes were highest among the elderly and younger generation in Spain when compared to UK, the proportion of elderly (75%) who received some form of care and support was similar in UK and Spain (Lowenstein and Daatland 2006). Further, care provision was higher among elderly who had poor health and functional limitations, among those who were single and older elderly (Lowenstein and Daatland 2006). Care provision to elderly parents by adult children were influenced by competing obligations among children, inter-generational geographic distance, social contact and reciprocity (Lowenstein and Daatland 2006). Further, while social norms governed care to elderly fathers, care to mothers were considered to be non-negotiable (Low1enstein and Daatland 2006).

This chapter uses descriptive evidence to examine the extent of social contact with adult children, living arrangements, morbidity, functional limitations, caregiving arrangements, economic support, and effect of care provision on caregivers, among elderly population living in an urban slum in Chennai. According to Census of India 2011, Tamil Nadu is the most urbanised state in India and 46.4% of the elderly live in urban areas in Tamil Nadu. In this study, care refers to social care comprising escort to health facilities, personal care assistance (ADL), assistance with household chores, help with medication adherence and running errands, and economic support involving transfer of financial resources between family members and from the state to elderly individuals. The main questions in the context of chronically deprived elderly that are addressed in chapter article are as follows: (1) What is the pattern of living arrangement among the elderly? (2) What is the extent of social contact between elderly and their adult children and other relations? (3) What is the level of self-reported morbidity, disability and activities of daily living? (4) What is the level of economic dependency? (5) What proportion of elderly receive care? (6) Who are the primary caregivers? (7) What is the effect of caregiving on the care provider?

Area of Study and Methodology

The research is based on primary data from a field survey conducted in 2005 in an urban slum within the Chennai Municipal Corporation, as a slum population can be considered to be a fair proxy for poor socio-economic status. Slums have been defined by the Statutory Section 3 of the Slum Areas (Improvement and Clearance) Act, 1956, as areas where buildings (Registrar General of India 2005, p. 417):

1. Are in any respect unfit for human habitation,
2. Are by reason of dilapidation, overcrowding, faulty arrangement and design of such buildings, narrowness or faulty arrangement of streets, lack of ventilation, light, sanitation, facilities or any combination of these factors which are detrimental to safety, health and morals.

On an average, a slum in Tamil Nadu had 196 households, with 356 households in notified slums and 129 households in non-notified slums (National Sample Survey Organisation 2003). Slums are characterised by overcrowding of poorly built tenements in a compact area with inadequate sanitary conditions (National Sample Survey Organisation 2003). While more than half of the tenements in notified slums are constructed with *pucca*[1] materials, more than half of the tenements in non-notified slums are constructed with *katcha*[2] and *semi-pucca*[3] materials. The use of pucca materials does not preclude from the structure being dilapidated, overcrowded and unhygienic. Sanitary facilities are particularly inadequate in non-notified slums, as majority of the tenements do not have latrine facilities and underground sewerage. Even in notified slums where 85% have latrine facilities, it is doubtful whether all the members living in a household will be able to access it, as research has shown that population densities were much higher in slum than non-slum areas of Chennai (Nagaraj and Majumdar 1982). The data also indicate that about 99% of the slums in Tamil Nadu had electricity connections to both households and streets. The government was the primary agent of garbage collection in notified slums in Tamil Nadu, with 99% of the slums reporting that garbage was collected by government agencies. Non-notified slums experience severe unhygienic conditions with nearly half of them reporting absence of garbage collection. As far as physical access to educational and health facilities is concerned, more than 90% of notified and non-notified slums have a primary school situated within one kilometre of the slum and more than half of the slums have a

---

[1]According to the National Sample Survey Organisation (2003:7), a pucca structure was one having walls and roofs made of 'pucca materials' which include cement, concrete, oven burnt bricks, hollow cement/ash bricks, stone, stone blocks, jack boards, iron, zinc or other metal sheets, timber, tiles, slate, corrugated iron, asbestos cement sheet, veneer, plywood, artificial wood of synthetic material and ploy vinyl chloride material.

[2]According to the National Sample Survey Organisation (2003:8), a katcha structure was one with walls and roof made of non-pucca materials like unburnt bricks, bamboo, grass, reeds, thatch.

[3]According to the National Sample Survey Organisation (2003:8), a semi-pucca structure was one which had either the walls or the roof, but not both, made of pucca materials.

government health facility within 1 km of the slum (National Sample Survey Organisation 2003).

A cluster of four slums—Thiruvethi Amman Koil Street, Manjakollai Street, Kathiravan Colony and Gajalakshmi Colony—was selected from Chennai Corporation's Zone 5 in Division 69, Ward 32, in Shenoy Nagar. The slums selected for the fieldwork were notified slums with about 1026 households. The slum population in Ward 32 had a very high proportion of SCs (62.8%). Information on some socio-demographic characteristics of the slum population in Chennai Municipal Corporation and Ward 32 of the Chennai Municipal Corporation is presented in Table 13.1.

The sex ratio of the slum population in Ward 32 is lower than that of slum population in Chennai Municipal Corporation. Further, slum population in Ward 32 has a very high proportion of scheduled castes (SCs) in comparison with that of the slum population in Chennai Municipal Corporation. It is clear that the slum population in Ward 32 is more disadvantaged in terms of access to education and paid employment. Earlier studies have brought out the high level of informal sector employment among the slum population in Chennai, which has been attributed to their low levels of educational attainment (Nagaraj and Majumdar 1982).

As part of the survey, which was conducted between July and October 2005, 206 elderly persons were interviewed on morbidity, treatment and caregiving. A complete census was conducted among the elderly in the slum cluster. However, those who did not wish to participate in the survey were not interviewed. The response rate to the survey was 73.3%, i.e. out of 281 elderly population in the slum, 206 participated in the survey (Table 13.2).

Phase 1 of the survey involved identification of the total elderly population in the slum. A complete house listing was conducted using a one-page interview schedule primarily to identify the total elderly population in the slum. The house listing was completed in three months, between December 2003 and February 2004. The total population enumerated in Division 69 in Chennai was 4494, and the elderly population was 281. Phase 2 of the survey involved conducting interviews with the elderly respondents on morbidity and treatment of ailments. The interview schedule, which was a structured schedule, draws on the schedule used by Susheela and Nagaraj (undated), Irudaya Rajan et al. (1999) and National Sample Survey Organisation (1998a, b).

The survey aimed to address the morbidity status and socio-economic conditions of the elderly respondents. The questions on household profile, which dealt with demographic and socio-economic indicators, were collected for all members of the household. The remaining sections were administered only to the elderly respondent. The household profile addressed issues like household composition, education, occupation, reason for withdrawal from the workforce and access to retirement benefits. The health profile contained questions about morbidity, medical treatment, utilisation of health facilities for outpatient treatment and hospitalisation,

Table 13.1 Socio-demographic characteristics of slum population in Chennai Municipal Corporation and Ward 32 of Chennai Municipal Corporation

Socio-demographic characteristics	Slum population	
	Chennai Municipal Corporation	Ward 32 Chennai Municipal Corporation
1. Total households	178,343	5251
2. Population		
Person	819,873	25,474
Male	415,296	13,040
Female	404,577	12,434
3. Sex ratio	974	954
4. Proportion of scheduled caste to total population (%)	32.8	62.8
5. Literacy (%)		
Person	76.3	67.4
Male	82.9	73.7
Female	69.6	60.7
6. Work Participation Rate (%)		
Person	33.6	29.1
Male	52.4	48.1
Female	14.2	9.1
7. Proportion of main workers to total worker population (%)		
Person	88.6	82.1
Male	90	83.5
Female	83.5	74.5
8. Industrial category of workers(%)		
Agricultural cultivators	1.2	1.3
Household labourers	0.5	0.5
Other industrial workers	2.3	1.9
Other workers	96	96.3

Source Registrar General of India, 2005, Slum Population Vol 1, Census of India 2001, New Delhi: Government of India

Table 13.2 Response rate in the survey

Gender	Number of aged population in the slum as per house listing	Number with whom questionnaire was canvassed in the survey			Response rate
		Direct	Proxy	Total	
Male	91 (32.4)	54	4	58	63.7
Female	190 (67.6)	147	1	148	77.9
Total	281 (100)	201	5	206	73.3

Note Figures in parenthesis are column percentages
Source Field survey, 2003–2005

expenditure on treatment, source of expenditure, caregiving, restricted activity and dietary restrictions caused by illness, and untreated ailments. Information on morbidity was conducted using a reference period of 30 days. Information was collected on the various types of disability, viz. visual, speech, hearing, locomotor and mental disability. The respondents were asked whether they used any disability aids. The schedule also tried to find out whether the elderly respondents could perform activities of daily living (ADL) like eating, bathing, dressing, using the toilet, transferring out of a bed or chair and walking, independently. The economic conditions of the elderly that were addressed were their source of income, type of pension that they got, economic dependence, ownership of assets (property), source of financial support and participation in self-help groups. Most of the concepts used are similar to those of National Sample Survey Organisation (1998a, b).

Socio-Demographic Characteristics of Elderly in the Slum

Table 13.3 presents information on socio-demographic characteristics of elderly respondents in the slum. There was a preponderance of elderly females (71.8%) than males (28.2%) and SCs constituted 46.1% of the elderly respondents. The proportion of young-old (59.7%) was higher than that of the old-old (40.3%). Nearly 60% of the elderly were widowed and 37.4% were married. Illiteracy among the elderly was high (60.7%), and work participation was less than 20%.

Living Arrangements and Household Headship Among the Elderly

Certain conditions like physically immobility, visual difficulties, diabetes, cardiac ailments, diarrhoea, fever require intensive care like assistance with ADL, cooking special foods and monitoring medicine compliance on a daily basis in addition to economic help, which might be feasible if there are resident family members. Shared dwellings as a preferred living arrangement could be the result of several factors like poor access to housing due to high costs, lack of means of economic support like own income, savings, retirement income, filial responsibility, inadequate social protection and disability. Besides the elderly, parent could be the owners of the house and adult children decide to co-reside with them owing to their inability to rent or purchase their own dwelling or due to separation from their spouse. Information on living arrangements which determines the availability of care and support shows that most of the elderly in the slum lived in shared households, despite the cramped dwellings/tenements (Table 13.4). Co-residence of the elderly with adult children was the dominant pattern of living arrangement, with 25.4% living with their adult sons and 18.2% living with their adult daughters. Nearly one-third of the elderly (32.4%) lived with their spouse and adult children. The elderly who lived alone

Table 13.3 Socio-demographic characteristics of elderly (as %)

Socio-demographic characteristics		Percentage distribution
Gender	Male	28.2 (58)
	Female	71.8 (148)
	Total	100 (206)
Caste	Scheduled caste	46.1 (95)
	Non-scheduled caste	53.9 (111)
	Total	100 (206)
Age group	Young old (60–69 year)	59.7 (123)
	Old (70+ year)	40.3 (83)
	Total	100 (206)
Marital status	Currently married	37.4 (77)
	Widowed	59.7 (123)
	Divorced/separated	2.9 (6)
	Total	100 (206)
Literacy	Illiterate	60.7 (125)
	Literate	39.3 (81)
	Total	100 (206)
Work	Worker	18.9 (39)
	Non-worker	81.1 (167)
	Total	100 (206)

Note Figures in parenthesis are actual numbers
Source Field survey, 2005

(8.5%) had comparatively higher ownership of property (45.8%) and close to 40% were economically independent (Balagopal 2009).

Figure 13.1 presents the information on household headship, which indicates that nearly half of the elderly (48%) lived in households headed by them, followed by households headed by adult children and spouse, which indicates that the elderly would play an important role in household decision-making. The relatively high level of household headship among the elderly points to ownership of the house and dependence of adult children on them for residential arrangements, which could enable reciprocal help like social care when the elderly fall sick. Thus, sharing residence with adult children could be a strategy adopted by the elderly to facilitate assistance in return.

Social Contact with Family Members

Social contact with adult children who live apart from elderly parents would lead to better emotional well-being among the elderly and also act as sources of extra-household social care and economic support. According to Lowenstein

Table 13.4 Percentage distribution of the elderly by living arrangements (as %)

S. No.	Living arrangement	Percentage distribution
1	Alone	8.5
2	Spouse only	7.1
3	Spouse and adult children	32.4
4	Adult son	25.4
5	Adult daughter	18.2
6	Other relations and non-relations	8.4
	Total	100

Notes The data for persons is a weighted average of the elderly male and female population available from the house listing
Source Field data, 2005

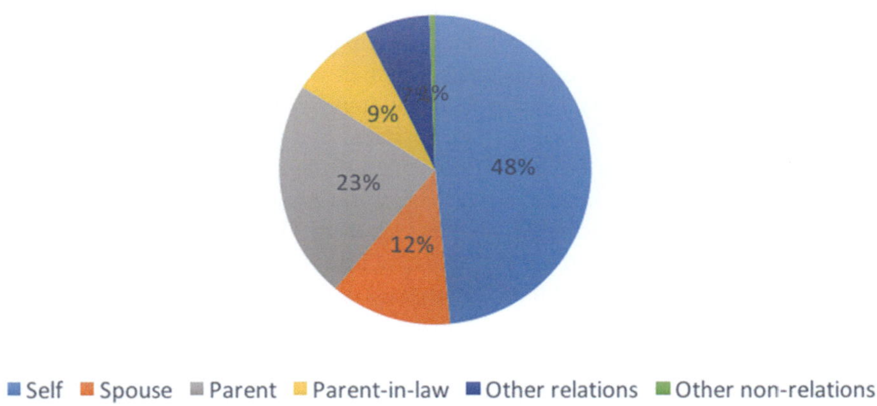

Fig. 13.1 Relationship of elderly to the head of the household (as %). *Notes* The data for persons is a weighted average of the elderly male and female population available from the house listing. *Source* Field data, 2005

(2005), women were more likely than men to maintain social relations between family members due to differences in the socialisation process between men and women. Social networks act as future support mechanisms and elderly women are known to have more diverse networks, comprising siblings, relatives, neighbours and friends, unlike men whose networks consist of work colleagues and immediate family (Grundy 2006). The information on social contact with children who live away from their parents shows that a little over half the respondents reported that their children visited them often and a little over one-third reported that their children visited them sometimes (Table 13.5). Around one-tenth of the elderly stated that their children had never visited them during the past 1 year. The data also indicate that for around half the respondents, sibling ties remained important. Strong sibling ties are known to substitute for the absence of spouse and adult children (Shanas 1979). A little over one-tenth of the respondents stated that their

Table 13.5 Social contact of the elderly with their family members who live elsewhere (as %)

Social contact	Frequency of social contact				
	Never	Sometimes	Often	Not recorded	Total
Adult children visiting elderly parents	11.1	35.0	53.4	0.5	100
Siblings visiting the elderly	38.5	42.5	14.0	5.0	100
Elderly visiting their relations	44.6	37.0	17.4	1.0	100

Notes The data for persons is a weighted average of the elderly male and female population available from the house listing
Source Field data, 2005

siblings visited them often and two-fifth of them said their siblings visited them sometimes. But more than one-third of the elderly stated that their siblings had never visited them during the past 1 year, which was due to siblings' old age, ill health and functional limitations, as well as conflict with siblings. When we look at the frequency of the elderly visiting their children and other relatives, a different pattern emerges. Only 17.4% of the elderly stated that they had visited their children and relatives often in the past 1 year and 37% reported that they had visited their children sometimes. Financial reasons, ill health and conflict with children were stated to be the reasons for not visiting them. The respondents appear to satisfy some of the dimensions of the solidarity framework, viz. availability of and frequent contact with adult children. Discussions with the respondents revealed that elderly women had a combination of horizontal and vertical networks when compared with men, with women having built up a strong network of relatives, friends and neighbours beyond the parent–child system.

Morbidity Among the Elderly

Although the elderly experienced high morbidity and disability, not all of them will require caregiving, as the nature of morbidity and severity of disability will influence the need for care. Functional limitations do not affect all the elderly and in some cases even if they experience difficulties in performing day-to-day activities, they have no choice but to do the work because of supply side constraints like availability and willingness of caregivers. Table 13.6 demonstrates that self-reported morbidity levels among elderly men and women in the slum were rather high going by the figures on proportion of persons who reported illness during a 30-day recall period. Further, 21.2% of the elderly reported some disability, indicating that some of the elderly suffer from morbidity and disability at the same time. However, disability was much higher among elderly men than women pointing to higher functional limitations and consequently higher care needs among men. Close to 60% of the elderly who were ill took outpatient treatment, suggesting

Table 13.6 Morbidity indicators among the elderly population

S. No.	Morbidity indicators	Male	Female	Person
1	Proportion of ailing persons (as %)	87.9	91.2	90.1
2	Disability (as %)	27.6	18.2	21.2
3	Outpatient treatment (as %)	58.9	59.8	59.5
4	Average number of illness episodes	1.3	1.6	1.5
5	Average duration of illness episodes (years)	5.5	4.9	5.1

Note The data for persons is a weighted average of the elderly male and female population available from the house listing; morbidity information is based on recall period of 1 month; the data do not add to 100 as proportions have been calculated for each type of indicator
Source Field survey, 2005

the need for escort to health facilities for some of them. The elderly suffered from more than one illness and this was higher among females than males. As information from National Sample Survey Organisation (2006) had shown that chronic disease prevalence was higher among the elderly than other age groups, it follows that care needs will also extend over a long period for those who had functional limitations. The average duration of illness among the elderly was 5.1 years, with males experiencing longer duration of illness than females.

According to Rice (1989), independence or dependence for activities of daily living (ADL) conditions the requirement for social, personal and supportive services. While the intensity of morbidity provides us with clues to the level of resources that will be required for the provision of geriatric care, it is functional limitations caused by the illnesses that indicate the quantum of need for physical care, which are primarily provided by the family. To understand the extent to which these illnesses cause difficulties, information was collected on ADL like eating, dressing, bathing, using the toilet, transferring out of bed/chair and walking without assistance from others. The literature on elder care has highlighted that it is these personal care tasks that are potential inter-generational stressors owing to the intimate nature of such caregiving responsibilities. Table 13.7 summarises the information on ADL for the elderly. It is evident that majority of the elderly who reported morbidity were able to carry out all the ADL without assistance and thus may not be care dependent. Activities that were performed with assistance from others to some degree were bathing (10.3%), dressing (8.2%) and using the toilet (7.7%). It is probably to avoid conflict causing situations created by high dependence on the family that some of elderly managed ADL by themselves despite experiencing difficulty in carrying out these tasks. Although 30–40% of the elderly experienced difficulty in using the toilet, transferring from the bed/chair and walking, they carried out these activities without assistance.

Table 13.7 Activities of Daily Living among elderly with illness (as %)

S. No.	Activities of Daily Living	Level of difficulty in performing activities of daily living among the elderly those with illness				
		Without assistance, without difficulty	Without assistance, with difficulty	With assistance	Not recorded	Total
1	Eating	93.2	2.6	2.6	1.6	100
2	Bathing	61.8	26.3	10.3	1.6	100
3	Dressing	75.1	15.1	8.2	1.6	100
4	Using the toilet	59.5	31.2	7.7	1.6	100
5	Transferring out of bed/Chair	49.2	43.2	6.0	1.6	100
6	Walking	46.8	47.9	3.7	1.6	100

Notes The data for persons is a weighted average of the elderly male and female population available from the house listing
Source Field survey, 2005

Economic Support and Social Care

Dependency among the elderly is socially created to some extent—not only by chronological age but also by inadequate government policies that have excluded large numbers of the poor from promotional and preventive social security. As a result of low levels of educational attainment and participation in low-paid employment in the unorganised sector and dependence on male wage in the case of most women, the poor enter old age with accumulated entitlement failures resulting in low levels of material resources. The added burden of illness, without the wide availability of free health care and poor reach of health insurance, creates dependency among the elderly for meeting their health care and daily living expenses. The choice of health service provider and nature of illness governs health expenditure. While treatment at government health facilities is free of cost in Tamil Nadu, it entails indirect costs like expenditure on transport and bribes paid to the staff. However, treatment at private health facilities involves considerable out-of-pocket expenditure. Dependence on the family is fraught with insecurities, particularly if the elderly live in households headed by their adult children, as the decision on whether to help and the quantum of support is made by children. Given adult children's employment in the informal sector or unemployment and financial commitments to their spouse and children, the amount of economic assistance might not match the elderly's requirements.

The question on economic independence uses the same definition as National Sample Survey Organisation (National Sample Survey Organisation 1998b). A person is considered economically independent if he or she does not require to take financial help from others in order to lead a normal life. The information on the

Table 13.8 State of economic independence of the elderly population (as %)

S. No.	State of economic independence	Male	Female	Percentage
1	Fully independent	50.0	18.2	28.4
2	Partially dependent	15.5	15.5	15.5
3	Fully dependent	34.5	66.3	56.1
	Total	100.0	100.0	100.0

Note The data for persons is a weighted average of the elderly male and female population available from the house listing
Source Field survey, 2005

level of economic independence reveals that more than half of the elderly respondents were fully dependent for economic support and this was pronounced in the case of elderly women, which reflects poor access to an independent source of income (Table 13.8). Importantly, despite living in a slum, more than one-fourth of the respondents were economically independent. Another 15.5% of the elderly stated that they were partially dependent.

We discuss two types of social care: one is escort to health facility and the other refers to care at home which includes help with household chores, medication adherence, assistance with ADL and running errands. While escort to health facilities would be needed only at specific intervals like on a fortnightly or monthly basis, care at home would be a daily exercise. It is evident from the data that most of the elderly were not care dependent; i.e., they were able to independently visit a health facility to take treatment and also manage household chores and manage medicine intake and personal care tasks (Table 13.9).

As families might not resent undertaking less frequent and less physically demanding care work like accompanying to the health facility, rather than the more strenuous and regular care tasks at home, a smaller proportion of elderly received care at home when compared to those who were escorted to a health facility. Other explanations for this pattern was that the elderly felt confident about handling their care needs independently or they downplayed their care needs due to the loss of

Table 13.9 Proportion of elderly who received social care (as %)

S. No.	Proportion of elderly who received social care	Male	Female	Person
1	Proportion of elderly who were escorted to health facility	56.1	41.1	45.9
2	Proportion of elderly with morbidity who received care at home	24.7	17.2	19.6

Note The data on social care are only for those elderly who reported morbidity and the data on economic support are for all the elderly; the data do not add to 100 as proportions have been calculated for each type of indicator; the data for persons is a weighted average of the elderly male and female population available from the house listing
Source Field survey, 2005

dignity and shame associated with being dependent on family members or their family members were unwilling to provide care. The issue of family members not providing care although the respondent needed help was evident in the case of a widowed elderly lady who had functional limitations and had lost control over her bowel movements. She co-resided with her son and daughter-in-law and was left unattended in her own faeces, as she was unable to perform ADL independently, especially using the toilet. During discussions, the daughter-in-law repeatedly shamed her mother-in-law about her incontinence and poor bowel control. There were frequent conflicts over this and her son and daughter-in-law had begun to neglect the elderly lady, which amounted to elder abuse.

The receipt of care is strongly gendered, with more elderly men than women being escorted to health facilities and receiving care at home when ill. We had earlier seen that a higher proportion of elderly men than women had some disability, which could partly explain the higher care for elderly men. Another factor that could be responsible for this pattern is that most of the elderly men were co-residing with their spouse who were their caregivers, as social norms dictate that wives should 'look after their husbands', whereas elderly women had no expectations that their husband would attend to their care needs.

Among the elderly who reported morbidity, the likelihood of the need for and hence receipt of care would be higher among those who experienced functional limitations like restrictions of their usual activities and those who were on bed rest during illness. All the elderly who were on bed rest on account of their illness received care at home and 86.4% were escorted to the health facility for treatment (Balagopal 2009). Among the elderly who had restricted activities during the illness, more than half (56.9%) were escorted to the health facility, but only around half of them received care at home, indicating that care was provided when illness caused severe limitations (Balagopal 2009).

Family Structure and Care

Living Arrangements and Care

Having a co-resident adult child increased the likelihood of the elderly receiving care, particularly if the spouse was also present, and in the case of single elderly among those who lived with adult daughters. Escort to health facilities and care at home was higher among those who lived with their 'spouse and adult children' and 'adult daughter' and least among those who lived 'alone' and 'only with their spouse' (Table 13.10). The receipt of social care being higher among elderly who lived with their spouse and adult children rather than with adult son indicates two possibilities. First possibility is that a large proportion of elderly men co-resided with their spouse and adult children, in which case most of their care needs were provided by the spouse. Second reason is that elderly women who co-resided with

Table 13.10 Social care and economic support by living arrangements (as %)

Living arrangements	Social care and economic support		
	Proportion escorted to health facility	Proportion who receive care at home	Proportion who receive economic support
Alone	3.8	4.5	62.8
Spouse only	23.3	8.0	75.1
Spouse and adult children	55.7	28.6	76.1
Adult son	26.9	12.3	88.7
Adult daughter	48.0	28.1	65.4
Other relations and non-relations	39.4	12.8	56.0

Note The data on social care are only for those elderly who reported morbidity and the data on economic support are for all the elderly; the data do not add to 100 as proportions have been calculated for each type of living arrangement; the data for persons is a weighted average of the elderly male and female population available from the house listing
Source Field survey, 2005

their spouse and adult children in which their spouse might be the head of the household, had higher chances of being cared for by her daughter-in-law, rather than if she were widowed and living as a dependent with her son and daughter-in-law. Social care was much lower among those who lived with their adult son when compared to those who lived with their adult daughter. Economic support was highest among those who lived with their 'adult son' and with their 'spouse and adult children' and was least among those who lived with 'others' and 'alone'. The elderly who lived alone were vulnerable in some aspects, as they reported high levels of morbidity and received the lowest levels of care and support. But, it must be noted that those who live alone were better resourced, as a large proportion of them owned their house and were economically independent, which was probably why they chose to live independently.

Demands on the Time of Family Members and Care

To what extent notions of filial obligations and reciprocity actually translate into care is governed by situational factors like competing responsibilities of the family (Lowenstein and Daatland 2006). The data on the number of working adults in a household and the number of children (0–14 years) will tell us to what extent employment outside the home and care responsibilities towards young children affect care and support to the elderly. Table 13.11 reveals that escort of elderly to the health facility was the least when there were no working adults in the household

Table 13.11 Proportion of elderly receiving caregiving by number of working adults and number of children below 14 years (as %)

Family characteristics		Social care and economic support		
Number of working adults		Proportion escorted to health facility	Proportion who receive care at home	Proportion who receive economic support
	0	40.4	17.3	66.8
	1	48.8	22.1	80.4
	More than 1	52.3	18.7	73.6
Number of children below 14 years	0	47.9	19.8	69.4
	1	38.8	19.0	79.5
	More than 1	39.5	19.8	65.9

Note The data on social care are only for those elderly who reported morbidity and the data on economic support are for all the elderly; the data do not add to 100 as proportions have been calculated for each type of family characteristic; the data for persons is a weighted average of the elderly male and female population available from the house listing
Source Field survey, 2005

and highest in households where there was more than one working adult. Care provision at home showed a decline when there was more than one working adult in the household and was higher in households in which only one adult member was in the workforce, and least in households in which there were no working adults. In slums, as the age at marriage is likely to be low, the family structure would be age-condensed. Consequently, the elderly and young children vie for the care-givers' attention. Escort to the health facility was highest when there were no children under 14 years in the household and when there was more than one child in the house probably because the older child could take care of the younger child or could even escort their grandparents to the health facility. An elderly woman reported that she felt bad to request her daughter-in-law to accompany her to the health facility, as the daughter-in-law had to take care of her young child. Care at home was highest in households with no children under 14 years and more than one child. In some instances, the elderly downplayed their care needs in order to avoid inconveniencing their families. Quite a few elderly women, who co-resided with their spouse, stated that they had postponed their cataract surgery, as there were no arrangements at home for help with household chores and they did not feel like requesting their daughters or daughters-in-law to come and stay over, as it would interfere with their children's schooling. Economic support to the elderly is highest in households with one working adult and one child (0–14 years) than in households with more than one child.

Source of Care and Support

Cultural conditioning determines who should provide care for whom. A study in Indonesia had shown that elderly men rarely took care of their wives (van Eeuwijk 2006). The caregiving pattern also reflects the strength and diversity of social networks among the elderly, which are known to differ by gender (Grundy 2006). Discussions with the respondents showed elderly men more often than not stated that filial obligations and reciprocity were reasons why adult sons should take care of them in old age. In the case of care from adult children and other relatives, the preference would be for the caregiver to be of the same gender, whereas in the case of spouse, especially elderly men, the caregiver was female. Information on care arrangements among the elderly shows considerable variations by gender and by the type of care, demonstrating the gendered division of household work (Table 13.12). Household duties and intimate personal care tasks are culturally conditioned to be the domain of women, which imposes a heavy burden on women who have to cope simultaneously with work and family responsibilities. Elderly men were mostly escorted to the health facility by their spouse (65.2%), whereas most elderly women were accompanied to the health facility by ‘other relations’ (31.6%) and adult daughter (24.6%). Very few elderly women were escorted to the health facility by their spouse (8.8%). The large share of ‘other relations’ who escorted elderly women to health facilities suggests that women had larger social networks than men, which were not restricted to the spouse and children. More than three-fourth of elderly men received care at home from their spouse, whereas none of the elderly women benefitted from spousal care at home. After spousal care, daughters were the next most important source of care at home for elderly men. Due to the intimate nature of care work, none of the elderly men reported receiving care at home from their daughter-in-law. These findings suggest that elderly men will face higher care deficits on death of their spouse, as adult children provide more care to their mother. Daughter-in-law (43.9%) and daughter (36.3%) provided most of the care at home for elderly women. The dominance of women in care work highlights gender-differentiated social roles in which men were not expected to perform care work.

Although elderly men benefitted more from social care, it was mainly from their spouse, whereas adult children provided more care to their elderly mothers than fathers. Informal discussions with the elderly respondents had revealed close emotional bonding between elderly mothers and their children. Further, adult children stated that it was the least they could do for their mother who gave birth to them and nurtured them.

Adult children, primarily sons, were the most important source of economic support, reinforcing their culturally conditioned role of providers of economic help. Economic transfers from the government were higher among elderly women than men, whereas more elderly men than women received economic help from

Table 13.12 Relationship of carer to elderly respondent by gender (as %)

Gender of elderly	Relationship of carer who escorts to health facility								
	Spouse	Daughter	Son	Daughter-in-law	Other relations	Non-relations	Government	Not recorded	Total
Male	65.2	13.0	8.8	–	13.0	–	–	–	100
Female	8.8	24.6	10.5	21.1	31.6	3.4	–	–	100
Person	26.8	20.9	9.9	14.3	25.7	2.4	–	–	100
	Relationship of carer at home								
Male	77.8	16.7	5.5	–	–	–	–	–	100
Female	–	36.6	2.4	43.9	17.1	–	–	–	100
Person	24.9	30.2	3.4	29.9	11.6	–	–	–	100
	Sources of economic support to the elderly								
Male	13.8	17.2	48.3	–	–	–	20.7	–	100
Female	11.6	16.5	34.7	–	5.8	–	30.6	0.8	100
Person	12.3	16.8	39.1	–	3.9	–	27.4	0.5	100

Note The data on social care are only for those elderly who reported morbidity and the data on economic support are for all the elderly; the data for persons is a weighted average of the elderly male and female population available from the house listing

Source Field survey, 2005

their sons. In 2005, when the survey was conducted, the state government provided Rs. 200 under the Old Age Pension (OAP) scheme.[4] In addition, the government also provided clothes twice a year to the elderly. The elderly was entitled to take part in the midday meal scheme in government and private-aided schools. While spouse was not the major source of economic support to the elderly, 13.8% of elderly men received support from their wife and 11.6% of elderly women received support from their husband, showing that elderly men receive more social care and economic support from their wife.

Effect on Caregiver

In the context of social care being highly gendered, multiple responsibilities faced by caregivers can cause stress and burn out. The data (Table 13.13) on effects of escort to the health facility on the caregiver reveal that caregiving was combined with other household work or paid work in most cases (63.9%), giving the caregiver no respite. In some cases, caregiving did impact household work and paid work outside the house. For 15.7% of the caregivers, providing help to their elderly relatives affected their household work. Another 12.1% of caregivers reported that they were not able to go to work the whole day on account of their caregiving duties. The adverse effect of care work on work participation of women results in economic and mental stress to the caregiver. The disproportionate burden of elder care on female caregivers can give rise to inadequate care provision to the elderly and in extreme cases can lead to conflict and violence against them. In the case of an elderly woman, when care interfered with paid work, she began to consider institutionalising her disabled elderly husband, as she was not able to care for him adequately. The potential for strained relations between the caregiver and the care recipient is immense, particularly when elderly suffer from incontinence and aggressive behaviour, as was reported by a woman who had to care for her father. According to the woman, '*My father troubles my mother who is visually impaired. He also displays abnormal behaviour. I had to give up my regular job because of him. He has no control over his bowels and he soils his clothes. I end up hitting him in frustration*'.

Discussion with the respondents revealed that conflict with family members was higher among those elderly who had to be escorted to a health facility than among those were independent. Further, care at home was non-existent among elderly men and women who reported frequent conflicts with their co-resident family members. Inter-personal conflict with family members was higher among those elderly who received economic support from their family, indicating that these could be the gateway to elder abuse.

[4]Currently, the government of Tamil Nadu provides Rs. 1000 a month to citizens eligible for state social protection schemes.

Table 13.13 Effect on caregivers' of care work (escort to health facility) work (as %)

Effect on caregivers work	Percentage distribution
Could not go to work the whole day	12.1
Could not go to work for part of the day	1.2
Could go to work only close to home	6.0
Had to keep aside household work to care for the patient	15.7
Had to continue with household work/paid work and care	63.9
Not recorded	1.1
Total	100.0

Source Field data, 2005

Conclusion

Although co-residence rates with immediate family members like spouse and adult children or with adult children, and frequent face-to-face contact with non-resident children was high among elderly respondents in the slum, it did not translate into high care provision for the elderly, as most were able to carry out ADL independently even if with some difficulty, suggesting a gap between need and receipt of care. The elderly received more care for less frequent and less physically demanding care work like escort to the health facility rather than the more strenuous and occasionally intimate care tasks at home, which had to be provided on a regular basis. Unlike social care, majority of the elderly received economic support. Care receipt showed variations by gender, with more elderly men than women receiving social care and more elderly women than men getting economic support. Among the elderly who received care and support, family involvement was substantial and the role of the state was significant in economic support (social security pensions) but absent in social care provision. Social care displayed gender asymmetry, with women providing bulk of care to the elderly, whereas most of the care recipients were elderly men. However, elderly women seem to be better resourced in terms of social networks to deal with old age, as they received more care from adult daughters, daughters-in-law and other relatives, unlike elderly men, whose predominant source of social care was their spouse. Adult sons, social security pensions and adult daughters were the most important sources of economic support. While notions of filial responsibility demand that adult children provide care for their elderly parents, actual implementation was influenced by situational factors such as competing demands on family's time because of participation in the workforce, childcare responsibilities and the nature of care that was required. Most caregivers reported that they had to perform care work along with household work and/or paid work, and discussions with them highlighted that some of them who had to care for incontinent parent/parent-in-law were considering institutional care, as they were unable to deal with the interruptions to their work and the resultant income shortfall. The gender asymmetry in caregiving has to be redistributed among other family members and the state, with a need to plan for formal, home-based care mechanisms for the elderly, in addition to universalisation of social security pensions for the elderly.

Acknowledgements Gayathri Balagopal is an independent researcher. This paper is based on the author's unpublished PhD Thesis entitled 'Morbidity, Medical Treatment, Social Care and Economic Support of the Elderly: A case study among the urban poor in Tamil Nadu State, India', 2009 (2009) at the Madras Institute of Development Studies, Chennai. The author wishes to acknowledge her PhD supervisor, Prof K Nagaraj and Reserve Bank of India for a fellowship during the PhD. This chapter is a modified version of a paper presented at a conference: 17th Biennial Conference of Association of Gerontology, India (AGI), and International Conference on Engaging and Empowering the Elderly (ICEEE-2014) during 15–16 September 2014, at Centre for Development Studies, Kerala.

References

Abe, A. K., (2010). *The changing shape of the care diamond: The case of child and elderly care in Japan, gender and development programme paper No. 9*. Geneva: UNRISD.

Balagopal, G. (2009). Unpublished PhD thesis entitled 'morbidity, medical treatment, social care and economic support of the elderly: A case study among the urban poor in Tamil Nadu State, India, Madras Institute of Development Studies, University of Madras.

Census of India. (2001). *Age data 2001*. New Delhi: Government of India. http://www.censusindia.net.

Census of India. (2011). *C-13 tables single year age data—(India/States/UTs/District) (Total, SC/ST)*. New Delhi: Government of India.

Eeuwijk, P. Van. (2006). Old-age vulnerability, ill-health and care support in urban areas of Indonesia. *Ageing and Society, 26,* 61–80.

Government of India. (2016). *Elderly in India: Profile and programmes*. New Delhi: Ministry of Statistics and Programme Implementation.

Grundy, E. (2006). Ageing and vulnerable elderly people: European perspectives. *Ageing and Society, 26,* 105–134.

Irudaya Rajan, S., Mishra, U. S., & Sarma, P. S. (1999). *India's elderly: Burden or challenge?*. New Delhi: Sage.

Irudaya Rajan, S., & Mathew, E. T. (2008). India. In S. Irudaya Rajan (Ed.), *Social security for the elderly: Experiences from South Asia* (pp. 39–106). New Delhi: Routledge.

Lowenstein, A. (2005). Global ageing and challenge to families. In M. L. Johnson, V. L. Bengston, P. G. Coleman & T. B. L. Kirkwood (Eds.), *The cambridge handbook of age and ageing* (pp.403–412). Cambridge: The Cambridge University Press.

Lowenstein, A., & Daatland S. A. (2006). Filial norms and family support in a comparative cross-national context: Evidence from the OASIS study. *Ageing and Society*, *26*, 203–223.

Nagaraj, K., & Majumdar, S. (1982). *Growth of Madras urban agglomeration, Madras institute of development studies working paper No.25*. Chennai: Madras Institute of Development Studies.

National Sample Survey Organisation. (1998a). *52nd round, morbidity and treatment of ailments, July 1995–June 1996*. New Delhi: Government of India.

National Sample Survey Organisation. (1998b). *52nd round, the aged in india: A socio-economic profile, July 1995–June 1996*. New Delhi: Government of India.

National Sample Survey Organisation. (2003). *58th round, 'condition of urban slums 2002'*. New Delhi: Government of India.

National Sample Survey Organisation. (2006). *60th round, "morbidity, health care and the condition of the aged"*. New Delhi: Government of India.

Palriwala, R., & Neetha, N. (2011). Stratified familialism: The care regime in India through the lens of childcare. *Dev Change., 42*(4), 1049–1078.

Peng, I. (2009). *The Political and Social Economy of Care: Republic of Korea Research Report 3*. Geneva: UNRISD.

Razavi, S. (2007). *The political and social economy of care in a development context: Conceptual issues, research questions and policy options, gender and development programme paper no. 3*. Geneva: UNRISD.

Registrar General of India. (2005). *Census of India 2001, Slum Population, Vol.1*. New Delhi: Government of India.

Rice, D. P. (1989). "Health and long-term care for the aged", papers and proceedings of the hundred and first annual meeting of the American economic association (May 1989). *The American Economic Review*, *2*(2), 343–348.

Shanas, E. S. (1979). The family as a social support system in old age. *The Gerontologist, 19*(2), 169–174.

Susheela, K., & Nagaraj, K. (undated). Health status and health seeking behaviour among agricultural labourers: A case study of a village in an advanced district in coastal Karnataka, Personal communication.

Chapter 14
Sibling Care Among Rural Elderly Widows

C. Aruna

Abstract Siblings of similar age have shared histories, similar concerns and feelings of solidarity. Intimacy and hostility are intertwined in the relationships, while they also complement and substitute parents and children in crisis. In the Indian context, a key to sibling relationships is the obligation of brothers to act as 'trustees' of their married sisters and replace friends in rural setting. In southern India, cross-cousin marriages and short marriage distance further cement their bonds and siblings of both genders constitute the core social world for women. But do such traditional patterns continue to operate where women claim equal right to property? To what extent does the support content and strength vary by gender of siblings? What constitutes care and intimacy in sibling relationships? These issues are less explored, and hence, the paper focuses on support transactions between elderly widows and their siblings and the intensity of role relations. The paper is inductive in nature and uses a qualitative methodology. Widows with siblings were interviewed through a semi-structured interview. Findings indicate strong ties between siblings, and brothers play a dominant role until children are married. Situations of strained relationships occur between brothers when expectations of widows are not being met. At later life, both brothers and sisters converge as confidants and companions. Sisters tend to form a thicker bond similar to friends, and brothers provide the ritual and service support. Results will be of particular interest to understand family relationships, elderly care and gender issues.

Keywords Siblings · Widowhood · Social networks · Social support

C. Aruna (✉)
Department of Sociology, Pondicherry University, Puducherry 605014, India
e-mail: aruna.chinnappan@gmail.com

S. Irudaya Rajan and G. Balagopal (eds.), *Elderly Care in India*,
DOI 10.1007/978-981-10-3439-8_14

Introduction

The demographic transition has altered the composition of Indian population with an absolute rise in aged population and in its relative proportion to the general population (Census of India 2011). Immediate family constitutes the primary caregiver, but the changing social milieu and family structure poses important challenges to the elderly and the care providers. The elderly are forced to mediate with the social adversities, changing life circumstances and social context (Cohen 2004; Dupertius et al. 2001). The life course perspective emphasizes the significance of 'convoys' in moderating the effects in later life (Kahn and Antonucci 1980). Convoys are a set of interpersonal relationships which are dynamic and lifelong while the changes may occur due to addition or deletion of members, role changes, social settings or the selectivity process. A reduced social interaction is noted in late adulthood (Carstensen 1995) and Gerontological literature in India also depict that network size in late life is comparatively lower (Chadha and Easwaramoorthy 1992; van Willigen et al. 1996). Many social relationships are a product of lifetime association with the potential to impact physical as well as psychological health (Kalavar and Jamuna 2008). Except a few findings, no systematic research exists regarding the social networks of elderly and the familial dynamics and social world of the elderly are less explained in the Indian context.

Family is a complex network of individuals with spousal subsystem, parent–child subsystem and the sibling subsystem (Minuchin 1974). The family systems theory maintains that what happens within any subsystem effects and is affected by events within other subsystems, while sibling subsystems are less explored in adult population. This 'forgotten relationship' has a major significance across the life span especially for women and the elderly. Sibling relationships are unique as they share a common cultural background, provide continuity for family sentiments and family belonging and have a prime structural positioning in the social world. The horizontal nature of ties characterizes equal power and sociability (Cumming and Schneider 1961) and hence closely parallels friendship ties, as cohorts experiences several life events and social cycles (Connidis 1992). Widows often feel closer to sisters due to the age similarity, events and encounters (Anderson 1984). Solidarity among siblings is bound by similarity and structural positioning, and differences occur due to competitiveness and interests (Teti 1992). These relationships are also characterized as close and rival (Ross and Milgram 1982), close and distant, and active and dormant (Walker et al. 2005). The shared history binds them, but lack of interdependencies in adulthood leaves a vacuum in the relationships during which other relationships such as spouse, children and in-laws take precedence. The sibling ties were usually heightened during divorce, widowhood, health issues etc., (Connidis 1992; Aruna et al. 2000) and form a dependable core group.

Siblings act as companions, provide emotional support, share reminiscence and validate each other's sense of self (Lee and Tallman 1980). They hold regard for each other, socialize through visits, telephone calls, sitting together, talking and discussing matters of mutual interest which are ordinary as opposed to exciting conversations (Scott and Roberto 1981; Allan 1977). Such mundane contacts maintain relationship (Adams 1968) and provide a sense of continuity over the life course, and confirm a sense of self in old age as persons still engaged in the world and as objects of affection and caring (Goetting 1986; Moss and Moss 1989; Ross and Millgram 1982). Therefore, siblings are identified to be of significant importance in later life, next to one's own nuclear family. The social network and societal norms set the parameters for sibling's location in each other's network. Such conditions determine the provision, degree and directness or indirectness in involvement of support transactions (Bedford 1995).

Sibling connections vary in the degree of involvement or indifference over the life span. Positive feelings are expressed in the presence of larger sibling relationships, while coalition exists within the groups indicating selectivity among siblings (Connidis 1989). Siblings closer in age share more common interests (Folwell et al. 1997) and having a proximate sibling increases communication among the entire sibling network and enhances general closeness to siblings. Women tend to outlive spouses, and many widowed sisters end up living together and share a special kind of closeness (McConville 1985). Siblings confide in siblings who are proximate (Campbell et al. 1999) and general preference for sisters prevail over brother (Wenger and Jerome 1999). Does it operate in Indian conditions too? Siblings are peers living into old ages and compose a large portion of the companionship network of single women (Connidis and Davies 1990). Siblings emerge as salient in the lives of women as well as among single and childless persons (Connidis 1992). The relationships are characterized by mutual availability. Bonds between aunts and uncles and their nieces and nephews tend to grow out of close sibling relationship (Wenger and Burnholt 2001). Siblings mediate relationship with other family members by being friends and surrogate parental figures (Ellington and Sotirin 2006). The voluntary nature of sibling ties makes it valuable. Does the intensity of sibling relationships vary by gender, class, proximity, marital status, parental status etc. and effect the amount and nature of sibling interaction? Given the importance of sibling connection for women, widowed and elderly, how does it operate in the Indian context?

Indian women are embedded in a conservative cultural milieu, which offers limited choices and greater constraints. The women are socialized for traditional stereotypic roles, and their mobility is bound by temporal and spatial limits. Siblings, especially brothers, are predominant sources of support for women due to their structural position and obligatory relationship across the life span. As in other cultural settings, women on widowhood often turn to siblings for support and services (Shanas 1979; Townsend 1957) and siblings act as role models for late-life transitions (Blieszner 1986; McGloshen and O'Bryant 1988).

Consanguines are bound by traditional obligations to daughters, and ritual links continue till the end of women's life in south Indian context. Hence, obligations find pre-eminence in the cultural idioms and in normative expectations. These usher siblings, especially male siblings, as prime supporters and their presence by itself has a major impact on the social well-being of widows (Aruna 2004). There is a normative obligation for brothers to act as 'trustees' of their married sisters, check on sisters and their children's well-being and extend gifts on important occasions. The short marriage distance, cross-cousin alliance and densely knit kin in rural community sustain the practice of traditional obligation, but also provides space for conflict (Aruna et al. 2001). A redefinition of traditional obligation is set in motion by many forces of change, largely due to women claiming 'equal right to property'. It manifests in gradual reduction in casual visits, in volunteering responsibilities and commitments to ritual transactions for young widows in agrarian setting (Aruna 2014). Seeking a share in the property in accordance to new laws and huge dowry at the time of marriage brings a change in traditional pattern of support with regard to quality and quantity of support. Women especially younger widows deem it a privilege to draw support from brothers, while male siblings feel it as a burden. On widowhood, brothers are the only relations who are approachable without barriers but strained relationships and snapping of ties reduces the potential support group of widows which extends well into old age. The spouses of siblings are vital players in moderating and the maintenance of relationship. How does this relationship pattern continue across the life span? Considering the potential fit of siblings and the limited studies in the context, important questions emerge as to whether the siblings draw closer together or drift apart over time. Are they considered important? If so, how are these connections meaningful? What are the structural dynamics in the network of siblings?

Methods

To understand the details of sibling connections, a qualitative approach was applied. In-depth interviews were conducted among 30 elderly rural widows who were identified through snowball technique. As the primary intention is to probe the changing social context, the rural region was identified to be fit, while sufficient knowledge and information regarding the cultural context through the earlier studies sets the base for the present study (Aruna 2004). Hence, elderly widows residing in farm houses of Pollachi Taluk, Coimbatore District, situated in western Tamil Nadu, India, were chosen for the study. Data Collection for the study was carried out during May and June 2015 for a period of 1 month. The elderly widows belonged to the Kongu Vellala Gounder community, a traditional agrarian peasant community, which is patriarchal and continues to follow the traditional norms. The lead questions focused on providers of care and support, structural positioning of siblings, their support transactions and the significance of such transactions.

Profile of the Sample

Widows in the study sample were aged between 61 and 85 years. The span of widowhood ranged from 2 to 15 years. All widows had children, and only six had unmarried children. Elderly widows with married sons co-reside with their family; in two cases, the widows lived in a partitioned portion of the same house and another four with married grand children who opted to stay in order to assist them. All the widows had land ownership which ranged between 1 and 20 acres, while financial transactions and agricultural management were largely handled by sons. Widows continued to be active in gender-based agrarian work, and except for minor health issues, none reported any major health problems. The consanguineal places of elderly widows were closer to the current place of residence, within a radius of 3–25 kms. All widows reported availability of at least one living sibling and had a maximum of nine siblings in compositions of brother–sister relationships and sister–sister relationships. Broadly, widows reported amicable relations with siblings but a few cases reported problematic relationships.

Widows of the study belonged to a generation where large 'sibships' were the norm and had combinations of biological siblings and step siblings due to the maternal mortality and the practice of polygamy prevalent 60–90 years ago. An age gap of 2–3 years were noted between siblings, and if the age gap between first and the last sibling was quite large, they were part of sibling care systems with eldest brother having greater responsibility. Older widows reported having more siblings and widows with larger subgroups reported demise of at least one sibling either due to old age or health issues. All siblings were geographically close though brothers were more proximate due to land holding while sisters were dispersed in different places within a radius of 30–40 kms. Being in contact with sibling group was quite common and was more frequent for larger sibling groups. Coalitions within sibling groups were also evident, and degree of closeness varied by virtue of proximity, emotional closeness, support transactions or 'ganging up' to defend other siblings.

Sibling Connections

The widows expressed that they had entered the later age of life and by agrarian standards '*pethu pithuru eduthachu*' meaning 'they have grand children'. Many of the widows whose children were currently married felt that their primary responsibility as a parent had been fulfilled and were relatively free and have moved into the stage of 'supporter' for the children's family. Widows with unmarried children were stressed and expressed their inability to conduct the marriage of their children, which would have been fulfilled if the spouse had been alive. These widows have greater expectations towards male siblings to support them, as they cannot find or finalize alliances themselves which needs networking, reinforcement and decision-making. The relatively settled widows though actively support children begin exploring spaces required for them and are more mobile (largely with the help of sons in two wheeler's) and assert themselves during which they find siblings and their family as a strong support group. These acts give an understanding that the widows and their siblings are in the process of drawing closer.

When examined how they would come together and for what purpose, the widows reported 'thicker descriptions' of the process. It begins with meeting more often, accompanying each other for common obligations, supporting each other's family when there is a need and staying with them when the elderly feel they need a change or when they 'feel down'. They describe being together in many ways such as '*ellaarum varuvanga*' (everybody comes), '*ellarum enkuda anbaa iruppanga*' (everybody will be affectionate towards me), '*nallathu kettathukku poiiruppom*'(for good and bad life cycle events we go and stay in siblings house), '*chenthu povom*' (accompany each other), '*vere enge poga mudiyum?*'(where else can we go and stay?), '*kooda poranthavangla intha jenmathala poranthachu, ini ethanai kalam irrupomo? sanda sacharuvu etharku?*' (we are sibling's in this birth, we do not know how long each of us will be alive, why should we have fights and arguments?), '*namakku avanga iruntha than mariyathai, illaina pasanga kuda mathikathu*' (we will have respect only if our sibling come and see us, otherwise even children would not respect us).

All elderly widows felt that they should be in contact with their siblings and felt that their presence matters, though four widows had strained relationship with brothers due to property disputes and for not gifting on ritual obligations. Strained relationships were more evident with brothers but not with sisters, and not every sister was considered with the same degree of closeness. Coalitions between siblings emerge to pressurize other dominant siblings for fair or due share of property settlement, and usually, younger brothers approach sisters for such settlement. This wedges the relationship, but usually, it happens at an earlier age and all siblings attempt to overcome such stress and strain. In the Indian context, rituals bind brothers and sisters and they have to mutually depend on each other but such compulsion does not exist between sisters. In addition, when elderly parents reside with son's, children (siblings) of both gender usually visit casually or during illness and with occurrence of parent's death, all children have a ritual role to perform, which bring siblings together and extended kin play a significant role to patch up. The ritual role of male siblings to sister's begin early right from the marriage of the sister as *kai korvai* (ritual of brother's symbolically offering hand to sister's husband) and sisters establish their privilege through performance of *innai seer* at the time of brothers marriage (taking a commitment from brothers through symbolic rituals that they will support sisters and provide all ritual entitlements). After the marriage as maternal uncle and maternal aunt, they occupy an important role in the sibling's families. The maternal uncle is seen as a friend, 'taken-for-granted relationship', potential father-in-law and emotionally close next to the mothers for sister's children. Male siblings have obligations in every life cycle ceremony of sister's family, and it continues up to the marriage of sister's children. Even on the death of a woman, brothers are expected to bring the '*porantha veettu kodi*' (ritual cloth from the house of birth) for the last rites of women, thus signifying the importance and link with consanguines for women in the south Indian agrarian context. The sibling relationships are strong and are made stronger through the community norms of conformity. They ridicule the brother or sister if they are unable to retain amicable contacts with each other and have repercussions in the social standing as well as in the future prospects of children, especially during marriage. When the sibling network is smaller, there is a greater pressure on male siblings and more demand from female siblings who indirectly vie on the property, which leads to break down of the strong, obligatory traditional ties and in larger sibling networks, expectations are only ritualistic. With changes occurring in the reduction of family size, siblings migrating to long distances and decline in celebration of life cycle ceremonies such dependencies have come down. With regard to sisters, there are no compulsions or social obligations towards other sisters, but each compete to receive the attention and support from brothers. At a later age, sisters become closer, meet and extend service support to each other and perform a functional role similar to friends.

Needs of the Elderly

The elderly widows spend most of their time on the farm and engage in household work and agricultural activities. Widows with married children limit themselves to work in female-oriented farm tasks and primarily take care of livestock and occasionally participate in household work if a demand arises. Less interpersonal problems prevail between the daughter-in-laws and mother-in-laws, and there is a clear segregation of work. Widows with unmarried children experience maximum problems and require support in many fields. They depend on others, especially male siblings, to find alliance and fix marriage for children, which have become increasingly difficult with rising age of the children. It becomes a major emotional stress for them, as they feel they are considered incapable to even find a match and are also report difficulty to continue with the household work due to the failing age and wait to excuse out of the binding household tasks. They look up to others for decision-making, purchase of goods and physical assistance in household maintenance. Added to these, they have to manage the middle-aged children whom they find increasingly difficult to handle and neither can shed responsibilities until the children are married. They are unable to even leave the household for a few days, as the children depend on them for cooking. Due to the widowhood, old age and unfulfilled marriage of children, they feel embarrassed to socialize and hence look for dependable and comfortable source of support among the close quarters and sisters emerge as immediate respite. These widows who are in need of high emotional and service support are the most vulnerable, as the problems are intertwined due to being a women, widow and an elderly. Other elderly widows are generally contended and interact with siblings for ordinary mundane needs such as socialization, satisfying needs of importance, concern, affection and companionship.

Nature of Relationship

Though contacts with a majority of siblings prevail, there is a strong difference between brothers and sisters and the degree of closeness. Elderly widows maintain a respectful and obligatory relationship with brothers, but a thin line of distance prevails, while the contact with sisters is more 'relaxed and comfortable'. With increasing age, widows reflect that 'problems, strain and conflict are part of everyday life and sustenance of relationships matter most'. Many widowed elderly maintain that if 'we don't have people coming to our house, especially siblings, we are looked down and do not have a social status'. This affects us as well as our children. No other person can have the same importance as our siblings, and we recognize it over a period of time. This relationship cannot be overlooked and has functional importance. If unavailable either due to strain or natural unavailability, it leaves a vacuum. Even in old age, brothers continue to provide the 'ritual needs' and fulfil 'obligations' to the extent possible and thereby give importance to the widow and her family. Women who had widowed early, i.e. those who had longer span of widowhood, indicated strained relationship with brothers, but those widowed later did not experience such strain in the relationships though unmet needs were expressed. Despite these, the widows had reconciled and renewed relationships when opportunity presented itself. Brothers also responded positively under pressure. To enable the maintenance of smooth relationship, apart from the sibling character, the nature of sibling's spouse and the children also matter. Lower age differences contribute to greater strain as well but less rigidity towards reconciliation. Thus, changing socio-cultural situation, life events and strain has not completely eroded the relationship of elderly widows with brothers who are a potential source of support, rivalry and conflict but a necessary contact.

Sisters are more often networked and occupy a prime place as women become older. Emotional closeness is more often cited with sisters, and widows express desire to meet them frequently unlike brothers whom they visit when there is a demand such as in situations of being unwell. Usually, sisters accompany each other to meet kin obligations such as attending marriage, death or similar events, which is a required need by the elderly. Apart from being with each other whenever there is a need, they mutually depend on each other on a voluntary basis while more closeness exists between children of female siblings. Sisters were reported to be docile at early ages, but when personal issues loom larger at later years, they share and confide with sisters especially the widowed elderly. Difficulties across the life span and greater freedom at later years foster greater sibling interaction. The common adage goes '*kolanthainga kadamaiya seivanga, annan thambi kitta urimai eduthakalam, akka thangachi than piriyama irruppaanga*' (children are obligated, with brothers we have a right and sisters are affectionate), '*perusa ethum panna maatanga aana kuda iruppaanga, nammala purichukuvanga*' (sisters may not be able to do much material transactions but they will always be with us and will understand us). Similar life situations and transitions enable closeness towards sisters and indicates sister–sister bond as thicker.

Support Transactions

Brothers have defined role in support transaction for widows. Brothers visit sisters once in a while, especially during occasions and give them gifts, and widows feel that the strong link between brothers itself is a support in all ages. Despite intact families, they do not want to sever ties with brothers. Presence of brothers in network gives widows a sense of security, and brothers are instrumental in handling crises in widow's family such as settlement of disputes with in-laws, neighbours or outsiders. Brothers often take a lead role to negotiate and settle on behalf of their sisters. In addition, brothers are prominent sources of support for fixing alliances for children to negate and negotiate the proposal of marriage and also remain an important anchor after marriage. Even in cases where children are problem creators, brother's step in to mediate and chide, as they are not considered 'outsiders'. Brothers are either direct providers of support or indirect through their children. They are potential members for widow's expansion of network, and some sibling's children get entangled deeper by being sons-in-law or daughters-in-law to widow's family, who are expected to provide qualitative support for the entire life span. Elderly widow's express feeling closer to daughters-in law who are their sibling's daughter and have an insider feeling towards them. Widows also receive respectable treatment in brother's family and hence feel privileged and honoured to visit brothers. No major material transactions occur at later age between siblings of either gender but casual visiting and stay over are often reported. At times when pending property disputes are closed, brothers gift sisters with small cash transactions in return for the signature as a token of appreciation.

Relationships with brothers are largely maintained at a ritualistic level though considered significant and important, and only a few are considered emotionally very close. Due to property disputes and quarrels, a few connections with male siblings have got snapped or strained. In connection with the claim to natal property, widows themselves are reluctant to take property from brothers as long as they are supportive and affectionate. They say '*naama ketta avanga kitta pogalaam, avanga ketta nammakitta varuvaangalaa?*' (if we go down in status we can go to them but if they go down in status can they come to us?). Such statements of widows give an understanding that if brothers are in a better economic position, it is a security and status for women and hence they contemplate on demanding their share. The widows are in an ambivalent stage and hence express that confining to traditional norms is better to maintain sibships. Unless changes in ritual transactions and demographic changes precede, exercising equal right to property has reservations. The women also report great survival difficulty quoting '*thulikku thuli thappuchu vaazhanum, athukku annan thambi venum*' (women have to escape from drop to drop to survive for which we need the support of brothers) who are a 'safety net' and provide a secure feeling and protection.

Widows are closer to sisters due to the care and bondage they share in old age. Whenever there is an emergency, female siblings immediately arrive and support and elderly widows also reciprocate similarly. Closeness is more associated with proximate and peer sisters.

In the Words of an Elderly Widow,

I live in a farmhouse, and normally, my social circle is limited to those who come to work in my field or visitors who occasionally visit me. I maintain myself according to the 'productive work', 'age-appropriate' discussions, advice and do not even engage in watching TV for a longer time and restrict to watching 'religious programs' or 'health programs' or sleep. What else can I do? I get some respite when somebody calls me especially my sister. She makes phone calls to me once in a week and enquires about my regular activities, health and other happenings. We visit each other at least once in a month or meet at some common functions. She is older to me and also a widow and had always been with me though we were not in talking terms for some time. It is only with her that I can discuss some personal issues which others do not empathize. If I am sick, she comes over and assists me in household work, sometimes we go shopping together, visit agricultural exhibitions, temples etc. We elderly widows generally do not go for movies or similar outings, but when my sister goes out along with her daughters and grand children, she insists that I join along with them. I also feel comfortable with them otherwise with whom can I go as my daughter stays at a distant place and cannot go with my son, as he may not feel comfortable. Generally, widows do not keep a 'bindhi' or 'pottu' on widowhood in our community, but my sister insisted that '*ean veru neththiyoda irrukera? santhana pottu vachikko*' (why should you have a bare forehead? Keep sandalwood bindhi) and she also bought one for me. I didn't keep 'bindhi' for a very long time since widowhood, and hence, initially I felt hesitant but she convinced how it is important and later I started applying whenever I went out with her. Now, I have got used to it. If my dressings are not good, she chides me saying 'why don't you dress up well?' She sometimes gets my clothes stitched, gets matching accessories, a good mobile pouch and once she also got 'a note book, pencil, pen, eraser etc.' and asked me to use it for maintaining accounts. She also came along with me once to select a good spectacle frame, and sometimes she also collects good reading articles for me from weekly magazines, newspaper etc. I feel good when she does such things. Can anybody else understand me so well? How do I define this relationship? It is a combination of friendship and sisterhood. In return, I also take some farm produce and spend time with her or be with her whenever required. My sister takes care of a temple and whenever she performs some pooja, she also expects me to be around. In one instance when my daughter was worried about me, I told her '*enakku periyamma irukiranga, ne kavalapadathe*' (do not worry about me, my sister is there).

Sisters are replacements of friends as lesser opportunity prevails in the rural community to have access to friends, and yet, such functional requirements are a necessity in all ages of life. Elderly widows are more comfortable with sisters to share painful life transitions especially widowhood, familial problems related to children, confidential matters, etc., and the age similarity enables them to share without any embarrassment. They also feel that sisters are able to fill the emptiness, understand, empathize, recognize their abilities, make them feel important, show concern and accept as them as they are, which the elderly widows express that no other role relationships can best fill such needs in a rural setting especially to those belonging to their generation. Thus, the sibling relationships provide stability, and become pre-eminent in old age which cannot be under estimated.

Conclusion

Indian rural elderly widows have a limited social world confined to the family subsystems. As they age, their social connections also change due to the absence of parental subsystem, reduction of sibling subsystem and expansion of children subsystem. Elderly widows feel more similar to sibling subsystem in terms of socio-demographic characteristics and resources. Widows embed well in sibling subsystems, but the stock of the years lived impinge on how the elderly blend into the system. Earlier studies on widows in agrarian community indicate pressure on sibling subsystem due to demands of widowhood and redefinition of obligations of brother. Several roller coaster changes occur in the lives of widows giving an impression that sibling relationship are under threat (Aruna 2004), but findings of the present study give a positive indication that the siblings' relationships draw closer and strained relationships gets renewed. Material transactions are kept aside, while emotional and companionship support gain major emphasis. Concern and recognition are valued by the elderly widows, and they draw comfort in the peer group of female siblings, which also supports findings of Anderson (1984). Brothers dominate in service support and are considered important for performance of ritual obligations. Sisters are considered the closest and have been revived after a period of dormancy. The age, gender, homogeneity and physical proximity provides the base for sharing and caring. Widows find siblings, especially sisters as an important functional group to meet the needs of elderly life. Among the elderly, the most vulnerable are widows with unmarried children or who have unfinished commitments of parental responsibility. They experience the multiple pressure of being a woman, widow and elderly and require support to re-socialize and adequate support flow from the social networks to meet the needs. The study draws findings from the widowed elderly women who have relatively large sibgroups who are activated even at a later age, but what happens to elderly widows with fewer or no siblings remains unanswered and demands attention from the academia and researchers. Results will be of particular interest to understand family relationships, elderly care and gender issues.

References

Adams, B. N. (1968). *Kinship in an urban setting*. Chicago: Markham.
Allan, G. (1977). Sibling solidarity. *Journal of Marriage and Family, 39,* 177–184.

Anderson, L. A. (1984). Intervention against loneliness in a group of elderly women—A process evaluation. *Human Relation, 37,* 295–310.

Aruna, C. (2014). *Social support of rural widows*. Chennai: Today Publications.

Aruna, C. (2004). Sibling relationships in widowhood. In P. A. Reddy (Ed.), *Problems of widows in India*. New Delhi: Sarup and Sons.

Aruna, C., Chandramohan Reddy, T., & Kaliappan, U. R. (2001). Negotiating with demands of widowhood and constraints of tradition in a farming community in western Tamil Nadu. *Social Science International, 17,* 17–24.

Aruna, C., Chandramohan Reddy, T., & Kaliappan, U. R. (2000). Patterns and processes of conflict with affines and consanguines—A measure of social well-being of rural widows in western Tamil Nadu. *Anthropologist, 2,* 153–159.

Bedford, V. H. (1995). Sibling relationships in middle and old age. In R. Blieszner & V. H. Bedford (Eds.), *Handbook of aging and the family*. Westport, CT: Greenwood Press.

Blieszner, R. (1986). Trends in family gerontology research. *Family Relations, 35,* 555–562.

Campbell, L. D., Connidis, I. A., & Davies, L. (1999). Sibling ties in later life: A social network analysis. *Journal of Family Issues, 20,* 114–148.

Carstensen, L. (1995). Evidence for a life-span theory of socioemotional selectivity. *Current Directions in Psychological Science, 4*(5), 151–162.

Census. (2011). Census report. (2011). Govt. of India.

Chadha, N. K., & Easwaramoorthy, M. (1992). Aging: Issues and rehabilitation strategies. In N. K. Chadha & S. Nath (Eds.), *Issues and trends in rehabilitation research*. Delhi: Friends Publication.

Cohen, S. (2004). Social relationships and health. *American Psychologist, 59,* 676–684.

Connidis, I. A. (1989). Siblings as friends in later life. *American Behavioral Scientist, 33,* 81–93.

Connidis, I. A. (1992). Life transitions and the adult sibling tie: A qualitative study. *Journal of Marriage and Family, 54,* 972–982.

Connidis, I. A., & Davies, L. (1990). Confidants and companions in later life: The place of family and friends. *Journals of Gerontology, 45,* 141–149.

Cummins, E., & Schneider, D. (1961). Sibling solidarity: A property of American kinship. *American Anthropologist, 63,* 498–507.

DuPertuis, L. L., Aldwin, C. M., & Bosse, R. (2001). Does the source of support matter for different health outcomes? *Journal of Aging and Health, 13,* 494–510.

Ellington, L. L., & Sotirin, P. J. (2006). Exploring young adults' perspectives on communication with aunts. *Journal of Social and Personal Relationships, 23,* 483–501.

Folwell, A. L., Chung, L. C., Nussbaum, J. F., Bethea, L. S., & Grant, J. A. (1997). Differential accounts of closeness in older adult sibling relationships. *Journal of Social and Personal Relationships, 14,* 843–849.

Goetting, A. (1986). The developmental tasks of siblingship over the life cycle. *Journal of Marriage and the Family, 48,* 703–714.

Kahn, R. L., & Antonucci, T. C. (1980). Convoys over the life course: attachment, roles and social support. In P. B. Baltes & O. Brim (Eds.), *Life-span development and behavior* (Vol. 3). New York: Academic Press.

Kalavar, J. M., & Jamuna, D. (2008). Interpersonal relationships of elderly in selected old age homes in urban India. *Interpersona, 2*(2), 193–215.

Lee, G., & Tallman, M. I. (1980). Sibling interaction and morale: The effects of family relations on older people. *Research on Aging, 2,* 367–391.

McConville, B. (1985). *Sisters: Love and conflict within the lifelong bond*. London: Pan books Limited.

McGloshen, T. H., & O'Bryant, S. L. (1988). The psychological well-being of older, recent widows. *Psychology of Women Quarterly, 12,* 99–116.

Moss, M. S., & Moss, S. Z. (1989). Death of the very old. In R. A. Kalish (Ed.), *Midlife loss*. Newbury Park, Calif: Sage.

Minuchin, S. (1974). *Families and family therapy*. Oxford, England: Harvard U. Press.

Ross, H. G., & Milgram, J. I. (1982). Important variables in adult sibling relationships: A qualitative study. In M. E. Lamb & B. Sutton-Smith (Eds.), *Sibling relationships: Their nature and significance across the lifespan*. Hillsdale, NJ: Lawrence Erlbaum.

Scott, J. P., & Roberto, K. A. (1981). *Sibling relationships in later life*. Paper presented at the annual meeting of the national council on family relations, Milwaukee, WI.

Shanas, E. (1979). The Family as a social support system in old age. *The Gerontologist,* 169–174.

Teti, D. M. (1992). Sibling interaction. In V. B. Van Hasselt & M. Hersen (Eds.), *Handbook of social development: A lifespan perspective*. New York: Plenum.

Townsend, P. (1957). *The family life of old people*. London: Routledge and Kegan Paul.

van Willigen, J., Chadha, N. K., & Kedia, S. (1996). Personal networks of older persons in a Delhi neighborhood: Power and the process of disengagement. *Indian Journal of Social Work, 57*(3), 414–428.

Walker, A. J., Allen, K. R., & Connidis, I. A. (2005). Theorizing and studying sibling ties in adulthood. In V. L. Bengston, A. C. Acock, K. R. Allen, P. Dilworth-Anderson, & D. M. Klein (Eds.), *Sourcebook of family theory and research* (pp. 167–190). Thousand Oaks, CA: Sage.

Wenger, G. C., & Burnholt, V. (2001). Differences over time in older people's relationships with children, grand children, nieces and nephews in rural North Wales. *Ageing and Society, 21,* 567–590.

Wenger, G. C., & Jerome, D. (1999). Change and stability in confidant relationships: Findings from the Bangor longitudinal study of ageing. *Journal of Aging Studies, 13,* 269–294.

Chapter 15
Migrant Children, Ageing Parents and Transnational Care Strategies: Experiences from Central Travancore, Kerala

Sreerupa

Abstract Ageing and transnational migration are two key processes of transformation in the contemporary world. Kerala has become the site for both these processes of transformation. While the state has had a history of large-scale transnational migration, which has had far-reaching influence on the economic and social fabric of the state, the ageing of its population has been a more recent phenomenon. Rapid demographic transition, population ageing and widespread transnational migration have come together to define the larger socio-demographic context of the state. Large-scale migration has dictated the absence of the younger generation of family caregivers, whereas an intensification of ageing in Kerala has resulted in a growing need for care for older persons staying behind. In this context, the paper discusses the implications of transnational migration of younger generation on the care for the older persons staying behind, examining the transnational care strategies and provisioning of care within the transnational families, in the empirical context of Kerala, which is at the forefront of both migration and ageing.

Keywords Ageing · Migration · Transnational care · Elder care · Kerala

Introduction

Ageing and transnational migration are two key processes of transformation in the contemporary world. Kerala has become the site for both these processes of transformation. While the state has had a history of large-scale transnational migration, which has had far-reaching influence on the economic and social fabric of the state, the ageing of its population has been a more recent phenomenon. Rapid demographic transition, population ageing and widespread transnational

---

Sreerupa (✉)
Centre for Development Studies, Prasanth Nagar, Medical College P.O, Ulloor, Thiruvananthapuram 695011, Kerala, India
e-mail: sreerupa.pillai@gmail.com

S. Irudaya Rajan and G. Balagopal (eds.), *Elderly Care in India*,
DOI 10.1007/978-981-10-3439-8_15

migration have come together to define the larger socio-demographic context of the state.

Large-scale migration has dictated the absence of the younger generation of family caregivers, whereas an intensification of ageing in Kerala has resulted in a growing need for care for older persons staying behind. In this context, the media reportage and public discourses have favoured an alarmist view of desertion of older persons in these transnational migrant families (see Knodel et al. 2010). They are portrayed as being left behind and uncared for by their migrant children. However, a growing stream of literature has documented that migration per se does not disrupt familial bonds and transnational families employ different strategies to overcome the distances, sustain bonds and also exchange care (Kilkey and Merla 2014; Parrenas 2005). Following this stream of literature, the paper discusses the implications of transnational migration of younger generation on the care for the older persons staying behind, examining the transnational care strategies and provisioning of care within the transnational families, in the empirical context of Kerala, which is at the forefront of both migration and ageing.

Transnational Migration and Population Ageing in Kerala

Kerala has a long tradition of migration, particularly international migration which has been rooted in a number of historical, economic and social factors. Migration from Kerala has seen several phases from the early twentieth century; however, transnational migration from Kerala has been dominated by movement to the Persian Gulf countries, a process that witnessed a boom in the 1970s (Zachariah et al. 2003). Later, over the years, migration flow diversified to newer destinations including the USA, UK, Europe, Australia, Canada.

Transnational migration from Kerala to the Persian Gulf countries, in particular, has been well researched and documented to have had a tremendous social, economic and cultural implication on the Kerala society (Gulati 1993; Zachariah and Rajan 2012; Rajan and Narayana 2010; Kannan and Hari 2002; Osella and Osella 2000). Interestingly, it has been argued that Kerala, hailed for its ‘model’ development[1] experience at low levels of income, has immensely benefitted through large-scale gulf migration and migrant remittances (Prakash 1998), described as the positive outcome of the Kerala model of development (Zachariah et al. 2003). In this context, much attention has been drawn to the migrant remittances received by the state over the years. It has been estimated that the total remittances received by the state was around Rs. 71,000 crore for the year 2014, nearly double the

[1]Since the 1970s, the Kerala’s development experience has received a lot of academic attention, for its accomplishments in terms of favourable statistical indicators of development in the field of health, education and other social and demographic indicators while remaining at low levels of income (see UN & CDS 1975; Franke and Chasin 2000; Raman 2010; Véron 2001; Kurien 1995 also see Thankappan and Valiathan 1998 for discussion on Kerala model of health).

remittances received back in 1998, (Zachariah and Rajan 2015) which comprises a striking 20% of the Net State Domestic Product (NSDP).

However, transnational migration and its influence across social groups were not as widespread as popularly believed; intense transnational migration from the state was in fact limited to smaller sections of the population. Zachariah and Rajan (2015) estimated that there were nearly 30 transnational migrants per 100 households in the state; however, these migrants were concentrated in only 19% of the households (Zachariah and Rajan 2015), with large differences across communities. For instance, Mar Thoma Syrian Christian community had the highest number of transnational migrants per 100 households at nearly 60 compared to only about 5 per 100 among the schedule caste and tribe communities (Zachariah 2016). Across Kerala, Syrian Christians[2] from the south-central[3] Kerala along with Muslims from the north and Ezhavas and Muslims (Mappila) from the south constituted the dominant emigrant communities (Zachariah et al. 2003; Kurien 2002). Notably among the Syrian Christians, both men and women migrated to find better paying jobs abroad which allowed them to migrate with their families and stay permanently in their host countries, compared to predominant temporary male migration among the Mappila and Ezhava communities (Kurien 2002). The large number of women migrants and more permanent settlement among the migrant Syrian Christian community had implications for the provision of elder care for those staying behind.

Apart from international migration, Kerala is also at the forefront of demographic transition. The state has the highest proportion of older persons (60 years and above) in the country, 12.6% elderly compared to only 8.6% at the national level, according to the 2011 census. Among the different communities in Kerala, Syrian Christians were also at the forefront of demographic transition. The Syrian Christians were quick to avail of the opportunities for birth control; as a result, fertility and mortality transition among them was more rapid than among the other communities (Zachariah 2001). About every fifth Syrian Christian in Kerala was aged 60 years or more, a far greater proportion than the state and national averages, making Syrian Christians one of the highly aged populations in the country (Zachariah 2001). By 2025, the proportion of older persons in the community is expected to touch 25% (Zachariah 2001).

---

[2]Syrian Christians are an ethnic religious group in Kerala. Syrian Christians popularly believe that they are the descendants of high-caste (Brahmin) converts of Thomas, the apostle of Christ, who was thought to have arrived in Malabar on his apostolic mission in AD 52 (Visvanathan 1993). They are high-status affluent minority community forming a small per cent of the population. The principal Syrian Christian denominations are namely, the Orthodox Syrian, Jacobite Syrian, Syro-Catholics and Mar Thoma Syrian.

[3]As per the Kerala Migration Survey, among all the districts in the state, Pathanamthitta has the highest proportion of Syrian Christians (Zachariah 2001) also has the third highest proportion of emigrants per 100 households at 43 after Malappuram and Kannur (Zachariah and Rajan 2015).

The study focuses on the experiences of Syrian Christians from the south-central Travancore in Kerala since they are at the forefront of both international migration and population ageing. The paper draws its empirical material from an ethnographic fieldwork conducted among 60 elderly households in Tirunadu, from 2008 to 2010, located in the high migration region of south-central Travancore in Kerala. Tirunadu[4] is a predominant Syrian Christian neighbourhood in a village in Pathanamthitta District. Tirunadu has been experiencing large-scale emigration of the younger population and their relocation abroad; the region has also been experiencing an intensification of ageing with the age structure being skewed towards older age groups. Tirunadu is located in the district that has one of the highest migration rates in the state as well as the highest proportion of elderly (21%) in 2004 (Zachariah et al. 2003; Rajan and Aliyar 2004).

Transnational Care Strategies and Practices

Till recently, migration was widely perceived to be disruptive since the family networks were thought to become geographically dispersed resulting in a weakening of social ties. However, researchers studying transnationalism have demonstrated that migration does not result in a radical break from the place of origin (Gardner 1995; Gamburd 2002; Zontini 2004). 'Transnationalism' research has shown that the lives of many individuals in the contemporary world increasingly transcend single localities and single nation (Levitt and Schiller 2004), and migrants forge and sustain connections between 'home' and 'away' and their family ties span across national boundaries (Glick Schiller et al. 1992; Gardner 1995; Gamburd 2002; Zontini 2004). Since these transnational social connections extend to the societies of origin, migrants as well as those staying behind are implicated in transnational families.

Bryceson and Vuorela have defined transnational families as 'families that live some or most of the time separated from each other, but yet hold together and create a feeling of collective welfare and unity', a process they term 'familyhood across national borders' (Bryceson and Vuorela 2002, pp. 3–7). Transnationalism, thus, challenges the static view of families based on the idea of co-residency and physical unity and forces us to take into account the possibility of spatial separation (Zontini 2004). There is a growing stream of literature which has documented that transnational families employ different strategies to overcome the distances, sustain bonds, maintain contact and exchange care (Kilkey and Merla 2014; Parrenas 2005). To analyse the provision of care within the transnational families, Kilkey and Merla (2014) offer a typology of caregiving arrangements within transnational

[4]Tirunadu is a social village, not an administrative, in the Pathanamthitta District. Although located in a rural area, it is far more urbanised. In Kerala, there is a rural urban continuum because of which the rural and urban areas are not drastically different from each other. The name of the place is a pseudonym, used to protect the privacy of the community.

families, which broadly represent different transnational care strategies. They categorise transnational family members involved in care exchanges according to certain spatial and temporal dimensions implicated in the care arrangement. The typology adopts a definition of care that goes beyond the provision of "hands on care" that involves activities such as feeding, washing or dressing a dependent family member, to include other forms of support proposed by the work of Finch and Mason (1993) like financial, emotional, practical support and accommodation (Kilkey and Merla 2014).

Under this typology, those transnational family members are categorised as 'Returners' and 'relocators' who move permanently in order to provide or receive care: the former are migrants who return to their country of origin at some point in the working and retirement life cycle; the latter are family members who move through the process of family reunification or sponsorship, to join their migrant kin in the receiving country. 'Visitors' are family members who travel to the host country to provide or receive care to or from their migrant relatives during short-term visits. 'Flying kin' circulate within the transnational family network to provide and/or receive care where and when it is needed. Finally, 'remainers' are migrants and family members who remain in their country of residence and exchange care transnationally. 'Reappearers' are migrants who provide or receive proximate care during short-term visits to their country of origin. Significantly, this typology is for analytical purposes and does not represent fixed categories, rather they vary over time in response to life course events, changing care needs and circumstances (Kilkey and Merla 2014).

Ethnographic fieldwork among the middle- and upper-middle-class migrant households in Tirunadu revealed that returning and relocating were a less common transnational care strategy, whereas a large majority remained in their countries of residence and exchanged care transnationally and/or during short visits. However, narratives of the older women revealed that earlier they often used to be 'returners', who used to return permanently to Tirunadu from where their husbands were working to look after their parents-in-law in their old age. I heard many accounts from older women (in their 80s and older) narrating how they returned to Tirunadu to look after their ailing parents-in-law or to give company to a widowed parent-in-law. Whereas, 'returners' were rare among the younger generation of women as most were working women, many of whom had migrated as the primary migrant, and had settled permanently with their families in the host countries. Since among the Syrian Christian's both men and women migrated to find better-paying jobs abroad, most often they migrated with their families and stayed on permanently in their host countries. The exception being those working in the gulf countries who were expected to return to Kerala at the end of their work life, due to restrictive government regulations for permanent assimilation into these countries.

On the other hand, 'relocators' were primarily older parents who were sponsored by their children to join them in the host country, to provide assistance with childcare and/or to receive company and care in their old age. Well-settled migrant children sponsored their parents' permanent relocation to the host countries, conditional to the immigration policies and regulations of the host nation.

Older parent's relocation was initiated by the migrant children particularly when one of the parents passed away and the other had to live alone. However, relocation was not always smooth, and there were numerous accounts of older parents finding their new life in a foreign country alien and choosing to return back to the village. 'Visitors' and 'flying kin' from Tirunadu were mainly grandparents, particularly grandmothers, who fly to take care of their grandchildren. It was common for grandmothers from both sides of the family (from fathers' and mothers' side) to take turns to provide childcare during regular short-duration visits of around six months each. Similarly, migrant children 'reappear' to mainly provide care to their ageing parents during short-term visits to Tirunadu, which involved routine visits as well as emergency visits. Most of the time, however, the older persons in Tirunadu and their children abroad were 'remainers' who exchanged care transnationally.

Significantly, more and more transnational families do not choose or are not able to choose reunion or return to address care needs of the older members of the transnational family and may choose to remain or reappear or visit and bridge the distances through their transnational care practices. The transnational elder care practices in Tirunadu include 'caring from a distance' and delegation of care locally. 'Caring from distance' commonly involves emotional caring and financial support. Emotional caring from a distance involves being in regular touch, staying connected, sharing of everyday lives, checking on each other, enquiring, advising, monitoring, basically recreating familiarity in a transnational space. This is achieved through long-distance communication practices such as ritualized, regular and longer phone calls, short quick calls or emergency calls, short message service (SMS), emails, regular or occasional Skype sessions and WhatsApp.

Similarly, financial support is provided from a distance in the form of occasional or regular remittances, joint bank account with the parent, salary cheques for care workers, bearing medical expenses, thoughtful 'cheque sent on each birthdays' and so on. Apart from these practices of caring from a distance, migrant children also purchase the services of local paid care workers to provide hands-on care services to the older parents. Further, visits provide an occasion for the migrant children to strengthen their emotional ties, provide hands-on care, organise and coordinate the paid care arrangements for the older parents staying behind.

Significantly, these contemporary migrant families are experiencing the distance of space and time very differently compared to the migrant families of the past, which has facilitated transnational care practices. Advances in the information and communication technology have made it easier and cheaper to stay connected 'anytime, anyplace and anywhere'. The rapid advance in the aviation and communication technologies around the globe has helped achieve a sense of 'time space compression'. As Levitt (2001, p. 22) has argued, new technologies have 'heightened the immediacy and frequency of migrants' contact with their sending communities and have allowed them to be actively involved in the everyday life in the sending communities in fundamentally different ways than in the past. Newer, cheaper and more efficient modes of communication (such as telephones calls, mobile calls, messaging, internet, emails, chats, video chats), transportation (air travel and fast trains) and banking (electronic fund transfers and ATMs) each have enabled migrants to remain 'connected' and maintain their lives both 'here and there' (Vertovec 2009; Licoppe 2004).

However, it is important to acknowledge that there are a host of challenges which these families face as they try to sustain family bonds and exchange care across nations. Distance and national borders are considered as the primary challenges facing the members of the transnational families (Zechner 2008). Distances could be a few thousand kilometres to thousands of miles, across national borders, across continents and across oceans. Distances affect mobility and result in more time required to meet simple care needs. Although advancement in the field of aviation has shrunk distances yet distances still affect the time and cost involved in visits and thus affect the number and frequency of visits back home. For instance, migrant children living in America, Canada or Australia may make less frequent visits back to Kerala like once in 3–4 years, while those in the Gulf countries may visit more frequently usually at least once every year.

The experience of transnational caring is further mediated by a host of other factors including resources, cultural norms, labour market integration, state policies and international regulations (Baldassar et al. 2007; Kilkey and Merla 2014; Zechner 2008). Different transnational families would have a different set of resources and capabilities to overcome distances such as economic resources, social network (Reynolds and Zontini 2006), access and familiarity with communication technology. Financial resources are very critical to make frequent visits, to sponsor visits and visas and to engage in frequent long-distance calls and other forms of communications. Similarly, access to means of communication and knowledge and ability to communicate using modern means of communication such as mobiles, emails, video chats become crucial to sustaining intimacy across distances. Moreover, how secure is the migrant children's position in the labour market, whether they have suitable visa's, how young their children are, how is the system of childcare and support in the country of residence, all these further add to the layers of complexity in transnational caring. In Tirunadu, the Syrian Christian migrant men and women's ability to exchange care transnationally and to achieve well-being of their dispersed family members comes from their privileged class and social position. The middle- and upper-middle-class Syrian Christian migrants' access to economic resources, strong social network, integration into the labour market and legal protection in the host country empowered them with a wider set of resources and capabilities to engage in transnational practices such as 'caring from a distance' as well as purchase of local paid care services.

Conclusion

Transnational families sustain and maintain family ties and exchange care across distances of space and time. However, akin to geographically proximate families, the intensity of the family ties and exchanges varies greatly even among the transnational families, ranging from simply keeping in touch to heavy exchanges of information, goods, money, people and emotions (Parrenas 2005). Some transnational families may try to eliminate distances by aspiring for geographical proximity to meet care obligations, while some others may try to bridge the distances and exchange care across distance. The practice of transnational care was, thus, not uniformly achievable by all social groups, since the experience of transnational caring is mediated by a host of factors including resources, migrants position in the labour market, integration into the host society, residential status, state policies and international regulations (Baldassar et al. 2007; Zechner 2008).

Acknowledgements This paper is based on the author's Ph.D thesis titled 'Transnational Linkages and Organisation of Care Work for the Aged: Experiences from Central Travancore, Kerala' submitted at the Centre for Development Studies of the Jawaharlal Nehru University, under the supervision of Prof. S. Irudaya Rajan.

References

Baldassar, L., Wilding, R., & Baldock, C. (2007). *Families caring across borders: Migration, ageing and transnational caregiving*. Hampshire: Palgrave Macmillan.

Bryceson, D. F., & Vuorela, U. (2002). Transnational families in the twenty-first century. In D. F. Bryceson & U. Vuorela (Eds.), *The transnational family: New European frontiers and global networks.* Oxford: Berg publishers.

Franke, R. W., & Chasin, B. H. (2000). Is the Kerala model sustainable? In G Parayil (Ed.) *Kerala: The development experience: Reflections on sustainability and replicability*. London: Zed Books.

Finch, J., & Mason, J. (1993). *Negotiating family responsibilities*. London: Routledge.

Gamburd, M. R. (2002). *Transnationalism and Sri Lanka's migrant housemaids: The kitchen spoon's handle*. New Delhi: Vistaar Publications.

Gardner, K. (1995). *Global migrants, local lives: Travel and transformation in rural Bangladesh*. Oxford, UK: Clarendon Press.

Glick Schiller, N., Basch, L., & Blanc-Szanton, C. (Eds.). (1992). *Towards a transnational perspective on migration: Race, class, ethnicity and nationalism reconsidered*. New York: New York Academy of Sciences.

Gulati, L. (1993). *In the absence of their men: The impact of male migration on women*. New Delhi: Sage Publications India Pvt Ltd.

Kannan, K. P., & Hari, K. S. (2002). Kerala's gulf connection: Emigration, remittances and their macroeconomic impact, 1972–2000. *Centre for Development Studies Working Paper No. 328*.

Kilkey, M., & Merla, L. (2014). Situating transnational families' care-giving arrangements: The role of institutional contexts. *Global Networks, 14*(2), 210–229.

Knodel, J., Kespichayawattana, J., Saengtienchai, C., & Wiwatwanich, S. (2010). How left behind are rural parents of migrant children? Evidence from Thailand. *Ageing and Society, 30*(05), 811–841.

Kurien, J. (1995). The Kerala model: Its central tendency and the outlier. *Social Scientist, 23*(1/3), 70–90.

Kurien, P. A. (2002). *Kaleidoscopic ethnicity: International migration and the reconstruction of community identities in India*. New Brunswick and New Jersey: Rutgers University Press.

Levitt, P. (2001). *The transnational villagers* (Vol. 46). Berkeley: University of California Press.

Levitt, P., & Glick Schiller, N. (2004). Conceptualizing simultaneity: A transnational social field perspective on society. *International Migration Review, 38*(3), 1002–1039.

Licoppe, C. (2004). 'Connected' presence: The emergence of a new repertoire for managing social relationships in a changing communication technoscape. *Environ Plan D Environment and Planning D: Society and Space, 22*(1), 135–156.

Osella, F., & Osella, C. (2000). Migration, money and masculinity in Kerala. *Journal of the Royal Anthropological Institute, 6*(1), 117–133.

Parrenas, R. (2005). Long distance intimacy: Class, gender and intergenerational relations between mothers and children in Filipino transnational families. *Global Networks, 5*(4), 317–336.

Prakash, B. A. (1998). Gulf migration and its economic impact: The Kerala experience. *Economic and Political Weekly, 33*(50), 3209–3213.

Rajan, S. I., & Narayana, D. (2010). The financial crisis in the Gulf and its impact on South Asian migrant workers. *Centre for Development Studies Working Paper No. 436.*

Rajan, S. I., & Aliyar, S. (2004). Gender ageing and social security. *Samyukta, women's initiatives, Trivandrum* (Vol. 4, No. 2, pp. 68–87).

Raman, K. R. (Ed.). (2010). *Development, democracy and the state: Critiquing the Kerala model of development.* London and New York: Routledge.

Reynolds, T., & Zontini, E. (2006). *A comparative study of care and provision across Caribbean and Italian transnational families.* London: London South Bank University.

Thankappan, K. R., & Valiathan, M. S. (1998). Country profile: India. Health at low cost—the Kerala model. *Lancet, 351*(9111), 1274–1275.

United Nations (UN) & Centre for Development Studies (CDS). (1975). *Poverty, unemployment and development policy: A case study of selected issues with reference to Kerala.* New York: Department of Economic and Social Affairs, United Nations.

Véron, R. (2001). The "new" Kerala model: Lessons for sustainable development. *World Development, 29*(4), 601–617.

Vertovec, S. (2009). *Transnationalism.* New York: Routledge.

Visvanathan, S. (1993). *The christians of Kerala: History, belief and ritual among the Yakoba.* Delhi, Bombay and Cacutta: Oxford University Press.

Zachariah, K. C. (2001). The Syrian christians of Kerala: demographic and socioeconomic transition in the twentieth century. *Centre for Development Studies Working Paper No. 322.*

Zachariah, K. C. (2016). Religious denominations of Kerala. *Centre for Development Studies Working Paper No. 468.*

Zachariah, K. C., & Rajan, S. I. (2012). *Kerala's Gulf connection, 1998–2011: Economic and social impact of migration.* New Delhi: Orient Blackswan.

Zachariah, K. C., & Rajan, S. I. (2015). Dynamics of emigration and remittances in Kerala: Results from the Kerala migration survey 2014. *Centre for Development Studies Working Paper No. 463.*

Zachariah, K. C., Mathew, E. T., & Rajan, S. I. (2003). *Dynamics of migration in Kerala: Dimensions, differentials, and consequences.* New Delhi: Orient Longman.

Zechner, M. (2008). Care of older persons in transnational settings. *Journal of Aging Studies, 22* (1), 32–44.

Zontini, E. (2004). *Italian families and social capital: Rituals and the provision of care in British–Italian transnational families.* Families & Social Capital ESRC Research Group, London South Bank University.

MIX
Papier aus verantwortungsvollen Quellen
Paper from responsible sources
FSC® C105338

If you have any concerns about our products,
you can contact us on
ProductSafety@springernature.com

In case Publisher is established outside the EU,
the EU authorized representative is:
Springer Nature Customer Service Center GmbH
Europaplatz 3, 69115 Heidelberg, Germany

Printed by Libri Plureos GmbH
in Hamburg, Germany